HANDBOOK OF QUALITY ASSURANCE FOR THE ANALYTICAL CHEMISTRY LABORATORY

SECOND EDITION

James P. Dux

VNR SPRINGER SCIENCE+BUSINESS MEDIA, LLC

Originally published by Van Nostrand Reinhold in 1990
MyCopy version of the original edition 1990

16 15 14 13 12 11 10 9 8 7 6 5 4 3 2

Library of Congress Cataloging in Publication Data

Dux, James P.
Handbook of quality assurance for the analytical chemistry laboratory / James P. Dux. — 2nd ed.
p. cm.
Includes bibliographical references (p.) and index.

DOI 10.1007/978-1-4757-5377-6
1. Chemical laboratories—Quality control. I. Title.
QD51.D88 1991 90-12577
543'.0028—dc20 CIP
www.springer.com/mycopy

HANDBOOK OF QUALITY ASSURANCE FOR THE ANALYTICAL CHEMISTRY LABORATORY

To my wife Catherine
and our children,
Ann and Thomas

Contents

Chapter 7.

Documentation for Quality Assurance 95

Chapter 8.

Organizing for Quality Assurance 117

Chapter 9.

Establishing a Quality Assurance Program 129

Preface to the Second Edition

When the first edition of this book was being written, the subject of analytical laboratory quality assurance was just beginning to be a topic of concern to analytical chemists. There was considerable confusion in the analytical laboratory community regarding just what constituted quality control and quality assurance, and what needed to be done in the laboratory to achieve these worthwhile goals. The literature on the subject was fragmented, and often represented a parochial point of view reflecting specific concerns of chemists in specific types of laboratories, working on specific types of materials. There were only a few books on the subject, and these also tended to reflect the backgrounds of the authors.

The intent of the first edition of this book was to attempt to bring some order and general principles to the field, and above all, to emphasize the practical aspects of laboratory management of quality assurance. Judging from comments received from readers, the book succeeded in these objectives reasonably well. However, in today's fast-paced world of science and technology, and in a field as "hot" as quality assurance in analytical chemistry, change is the order of the day. Therefore I was intrigued when the publisher suggested a second edition might be in order, and readily agreed. Although the basic principles remain the same, discussions with analysts, laboratory supervisors, and managers indicated many areas where improvements could be made.

For example, new chapters have been added on sampling and quality assurance; laboratory facilities and quality assurance; and auditing for quality assurance. Very little of the first edition has been discarded, but many topics have been expanded considerably. The chapter on computers has been completely rewritten in view of the rapid changes in that field. The chapter in the first edition on planning and organizing for quality assurance has been split into two chapters, one on planning for quality assurance and the other on organizing and establishing a quality assurance program, and new material on mandated quality assurance programs has been combined with the material on laboratory accreditation. Numerous examples, especially those involving mathematical calculations, have been added at the suggestion of some readers. In short, this edition is very nearly a new book, and I can only hope it is as well received as the first edition.

CHAPTER

1

Quality, Quality Control, and Quality Assurance

One of the strongest trends in modern society is the continuing evolution from a manufacturing to a service-oriented economy. The desirability of this may be arguable, but the fact remains that a smaller and smaller percentage of the work force is directly involved in manufacturing goods, and an increasingly larger percentage is concerned with providing services. Another strong trend is the evolution toward what has been called an "information society." Information is now recognized as a fourth resource, along with capital, labor, and raw materials as the bases of a dynamic economy.

The analytical chemistry laboratory, except for academic and research laboratories, fits very neatly into this analysis. The sole reason for the existence of analytical chemistry laboratories is that they provide a service to other segments of society, namely information regarding the chemical identity and composition of the various materials society uses.

Information, considered as a resource, has certain characteristics which need to be considered. For example:

Accuracy. Information must be accurate, or it is useless. There is an old maxim "no data is better than poor data." Poor data may lead the recipient into making poor or disastrous decisions, which may never be made in the absence of data. Misinformation or disinformation may be of value in the world of "cloak and dagger," but not in the everyday world.

Timeliness. Information often must be obtained in time to make the correct decision. If not, the information may be useless.

Cost. Information is seldom free.

Information must satisfy the purchaser's needs, or it is useless.

One intriguing characteristic of information is that it may be sold or given away, yet it is still retained by the seller or donor. This has important implications for computer software manufacturers and publishers in these days of personal computers, photocopy machines, and facsimile machines.

Another characteristic of information is that it can usually be easily and inexpensively transferred, especially in these days of relatively inexpensive telecommunications.

Looking at these characteristics it is easy to see that they all apply to analytical data. Analytical results must be accurate and timely, can be costly, must satisfy users' needs, can be sold or given away and still be retained, and can be quickly and inexpensively transferred. Our major concern in most of this book will be with the first characteristic, the accuracy of the data, although other characteristics will be discussed.

Based on this discussion of analytical data as information it follows that quality assurance of analytical data is important because:

Analytical data are often critical in the expenditure of large amounts of money. Decisions on major projects often are based on data coming out of the laboratory.

Accurate analytical data are often important to public health and safety.

Analytical data are important in evaluating, monitoring, and protecting the environment.

Analytical data are often critical in litigation, and in today's litigious world every organization and many individuals are vulnerable to lawsuits.

Analytical data are often quite expensive. Today's laboratory instrumentation and equipment are costly, and when the price of something rises concern about quality increases.

Thus there are four categories of people concerned about the quality of laboratory data. First, there are the laboratory's "clients," that is, those who use the laboratory's data and who, in the long run, pay for the laboratory's existence. Second, there are the outsiders who evaluate the laboratory's data, for example, government agencies for compliance to regulations, or lawyers involved in litigation. Third, there is senior management within the laboratory's parent organization who have a legitimate concern that the money to operate the laboratory is being wisely spent, and finally, there are the laboratory workers themselves who certainly have a concern that the work they do is of the highest professional quality.

However, discussions with analytical chemists and laboratory managers reveal that there is much confusion in the laboratory community regarding the meanings of these terms "quality control," "quality assurance," and even "quality." In particular many chemists are confused regarding the difference between quality control and quality assurance, and many appear to believe that there is no difference and the two terms can be used interchangeably. Unfortunately, the literature is not much help. If one reads five different papers on quality assurance one will be likely to find five different definitions of the subject. They will be similar, but subtly different, and often different enough to be contradictory. Before proceeding further then, we will define these terms, at least as they will be used in this book.

QUALITY

The word "quality" is one of those portmanteau words that have many different meanings, depending on the context. Webster's Seventh New Collegiate Dictionary (1969) offers five major definitions, with about fifteen subdefinitions. When used as a noun and applied to material objects it is a term that we assume everyone understands, but this is not always true. For example, one definition is that quality is the degree to which the object satisfies the needs of the user. At first glance this seems reasonable, but when examined further it is not, since the needs of the user are often ill-defined. For example, if a user needs to travel from point *A* to point *B*, a Ford Escort will do the job as well as a Porsche, but few would argue that the cars are of equal quality. One can add other user needs, such as low cost, comfort, safety, and freedom from repairs, but even this will not remove the ambiguity.

If one attempts to apply this definition to an analytical result, that is, that the quality of the result is the degree to which it meets the user's needs, the same kind of ambiguity results: Users frequently do not know what their needs are, nor can they accurately define them.

Fortunately, there is one definition of the quality of an analytical result that is relatively unambiguous, scientifically correct, and easy to understand. This is that the quality of a result is equivalent to its accuracy, that is, the degree to which the result approaches the "true value" of the thing being measured or determined. (Some persons may quibble with this definition, arguing that accuracy is an absolute, a result is either accurate or not, but we will ignore them in this book.) Unfortunately, the true value of the analyte is seldom known. However, by analogy with results obtained on materials of known composition we can at least estimate the accuracy of a result on an unknown.

One point must be made regarding the definition of accuracy in the above definition of quality. In the older literature the word "accuracy" is often equated with freedom from bias, or constant error. The analogy frequently used is that of a marksman shooting at a target. If the resulting shots are widely scattered but average on the bull's-eye the marksman is said to be accurate, while a similar series of shots that are closely grouped but whose average is far from the bull's-eye are called precise, but inaccurate. In this book we will define accuracy as the degree to which the results approach the true value whether the difference is due to bias or random error.[1]

QUALITY CONTROL

If quality is defined as equivalent to accuracy, then quality control must relate to accuracy control. (In fact, we could speak of accuracy, accuracy control, and even accuracy assurance and life would be a little easier, but unfortunately historical precedent is against us.) The following is the definition of quality control that will be used in this book:

> Those laboratory operations whose objective is to ensure that the data generated by the laboratory are of known accuracy to some stated, quantitative degree of probability.

What this definition means is that the laboratory should be able to attach to every result reported a range within which the true value of the analyte can be stated to lie with a certain degree of confidence in the probability of being correct (the confidence limits of the analysis). This does not mean that the confidence limit or precision of the analysis *must* be reported with the result (although this might be a good idea), but rather that the laboratory should be able to make this statement, if required.

The kinds of operations referred to in this definition are those that are generally applied as good scientific practice: instrument calibration, personnel training, use of pure reagents, use of standards and reference materials, and so on.

In many laboratories, at least those without a good quality control program, if a chemist or supervisor is queried about the accuracy of a given result, data will be supplied from the literature, or laboratory data that were taken some time ago by laboratory personnel. These data generally refer to the "quality" of the method used and not to a specific result. They are often obtained by asking one of the better, more experienced analysts, to evaluate the accuracy and precision of the method. He or she analyzes a series of reference samples of known concentration, using freshly made reagents, calibrated instruments, and careful techniques, and then calculates bias and standard deviation. This is then accepted as being applicable to all future analyses using this method.

However, consider the run-of-the-mill sample analyzed in the laboratory by this same method. The analysts are likely to be less experienced, reagents are made up incorrectly, instruments may be calibrated only infrequently, and so on. In general, the run-of-the-mill sample quality will be much worse than that obtained in the first case.

The point of this is that quality control is more than just a one-time evaluation of the method used. It is instructive to ask what is being controlled, when speaking of quality control. The answer is the analytical system. Every analytical chemistry laboratory can be considered to be a system for generating analytical results. The system consists of all those things that can have an impact on the results: personnel, reagents, reagent solutions, methods, instruments, equipment, and even the laboratory environment and facilities. Later chapters in this book will discuss means of monitoring, evaluating, and controlling the analytical system.

It is also important to note that, in the above definition, nothing is said about the *degree* of accuracy or quality of the data. All measurement processes, including analytical chemistry, are subject to errors. The objective of quality control is not to eliminate, or even minimize errors, but to measure or, more correctly, to estimate, what they are in the system as it exists. There are many reasons for choosing one analytical method, instrument, or system rather than another besides accuracy or quality of results, for example, cost, time, convenience, or safety. Improving quality of data is a worthwhile goal, but it is not a function of quality control.

QUALITY ASSURANCE

Having defined quality and quality control we now come to quality assurance. The word "assurance" implies the ability to demonstrate or prove something to someone else. Thus quality assurance refers to the ability of the laboratory to demonstrate or prove to someone not working in the laboratory that the quality of the data is what the laboratory says it is. The person being assured may be a client or user of the laboratory data, senior management who are paying for good quality, a government agency investigating compliance with regulations, or a lawyer or judge in a legal dispute. As might be expected, quality assurance relies heavily on documentation.

The following definition is offered for quality assurance as applied to laboratory operation:

> Those laboratory operations undertaken to achieve the following objectives of documentation:
>
> 1. That quality control procedures are indeed being implemented in the laboratory. An example is quality control requires that all pH meters be calibrated before use. Quality assurance requires that the fact that the meter was calibrated, and the results obtained, shall be recorded.
> 2. Assure that the accountability of the data is maintained. "Accountability" in this context means that the data reported do, in fact, reflect the sample as it was received in the laboratory, that is, that sample mix-up was avoided and the sample was properly preserved prior to analysis.[2]
> 3. Facilitate traceability of an analytical result. Every result reported should be traceable to the date of analysis, the analyst(s) who ran the sample, the method used, the instrument(s) used and its (their) condition, and the status of the analytical system at the time of analysis.
> 4. Ensure that reasonable precautions are taken to protect the raw data against loss, damage, theft, or tampering.

Note that in this definition quality control is mentioned only in the first objective. Quality control is a subtopic under quality assurance. It is the most important and is the core activity under quality assurance, but the other objectives are also important. This is the point of most confusion in the scientific community over the two terms quality assurance and quality control. Most chemists and their supervisors are sympathetic to the practice of good quality control, since this is in essence merely good scientific practice, but the necessity for quality assurance is often not perceived as important and is often ignored.

QUALITY ASSURANCE AND LABORATORY MANAGEMENT

If the analytical laboratory existed in a vacuum, generating data purely for internal use or as an intellectual amusement, there would be no need for quality assurance, and quality control would be sufficient. However, this is almost never the case, except for academic laboratories in a teaching mode. (Perhaps this is why the subject of quality assurance is not taught in academia.)

As mentioned at the beginning of this chapter, the analytical laboratory must interface with the rest of the world in a myriad of ways. The data generated are of intense interest in many cases to persons outside of the laboratory, and they are going to be interested in the ability of laboratory management to prove or assure them of the data's quality. In fact, one definition of quality assurance is the procedures that enable the laboratory to produce "legally defensible" data. While this is a good definition, the one given above is preferred because it spells out in greater detail how to go about doing this.

Another disadvantage of the legally defensible definition is that it emphasizes the legal aspects and many, if not most, chemists will never have to defend their data in a court of law. Therefore, the reasoning goes, why bother with quality assurance? One answer to this is that, in today's litigious society, no one can say that he or she will never be in a court of law defending his or her work. Another answer is that, whether or not litigation is involved, as a practical matter all analytical chemists and laboratories are continually subjected to controversy over the validity of results and a good quality assurance program can be of great value in resolving conflict.

Since the establishment and maintenance of a good quality assurance program requires allocation of human and financial resources, it is clearly a management responsibility. Unless laboratory management is clearly committed to quality assurance it will not happen. Laboratory workers on their own initiative will not establish quality assurance procedures, although they will practice quality control as a mark of professional pride. However, quality assurance requires more work, especially paperwork, which very few people willingly undertake unless it is clearly required to maintain their employment.

The degree of rigor of the quality assurance program in any given laboratory will also be a management responsibility. This will vary with the laboratory's overall mission. A laboratory engaged in environmental work for compliance with government regulations will require a rigorous program to ensure acceptance of their data. Crime laboratories whose data are often involved in legal disputes are another example of the need for rigorous quality assurance. Independent testing laboratories may need a rigorous program to defend their data and also as a desirable marketing tool. On the other hand a small quality control laboratory attached to a manufacturing site, whose data seldom are scrutinized outside of the parent organization, may not need a very rigorous program.

Management must decide the level of quality assurance because quality assurance is a cost item, and also because ultimate responsibility for data quality rests with management. One often sees or hears the statement "quality assurance is free" meaning that the benefits outweigh the costs. While the author certainly believes this, it is not usually easy to establish, at least quantitatively in terms of the "bottom line." The reason is that while the cost of a quality assurance program is often easy to establish with routine accounting and labor allocation procedures, many of the benefits are intangible.

The benefits of a good quality assurance program are analogous to those of a good safety program. In both cases, the important benefits are those of disasters averted, that is, the accidents that don't happen or whose consequences are minimized in the case of the safety program, and the samples that don't need to be rerun and the controversies avoided in the case of the quality assurance program.

One intangible benefit, which the author has observed in every laboratory with a good quality assurance program, is an increase in employee morale and pride in the work. One reason for this is that the program provides a constant reinforcement of the feeling that the work is being done properly. With a good quality assurance program the analyst not only knows the quality of the data being produced, but can prove it if need be. The ultimate vindication of the quality assurance program comes when analysts must defend their results to a client or some other outside agency, or even in a court of law.

REFERENCES

1. Kirchmer, C. J. Quality Control in Water Analysis, *Environmental Science and Technology*, 17:174A–184A (1983).

2. Kanzelmeyer, J. H. Quality Control for Analytical Methods, *ASTM Standardization News*, pp. 22–28 (October 1977).

CHAPTER

2

Error and Statistics

In Chapter 1 quality control was defined in terms of the operations needed to estimate the accuracy of an analytical result, that is, the degree to which the measured result approaches the true value of the thing measured. We could just as easily have defined it in terms of the error in the result, since the error is the inverse of the accuracy. The larger the error, the smaller is the accuracy.

ERROR

Errors in measurement can be divided into three categories: errors due to human error, commonly called "human errors" or "blunders," errors due to bias, and those due to random causes.

Blunders

Blunders or human errors are the most insidious type of error to attempt to control. They are intermittent, occur at random, and may be small or so large as to result in totally inaccurate results. Examples are misreading instruments or equipment, transposing digits when data are recorded or transferred from one piece of paper to another, misreading written analytical methods, inputting incorrect data into a computer or calculator, and a myriad of other things humans can do which can impact on the result of an analysis. "To err is human" as the proverb says, and as long as we have human beings working in laboratories mistakes will be made.

This fact is one of the driving forces leading to the computerization of laboratories, but, as is well known, computer results are only as good as the data input. Studies made before computers were laboratory tools, comparing derived results with notebook raw data, indicated that about 1 percent of the results were in error due to calculation errors. Computers will eliminate calculation errors, but cannot eliminate errors due to incorrect data input.

Most blunders are impossible to control and correct, short of running every analysis twice with two different analysts, unless the results of the analysis are clearly out of line. However, there are techniques for identifying "blunder prone" individuals, which will be discussed in later chapters (cf. Chapter 3, Blind Samples).

Errors Due to Bias

These errors are often called "assignable errors" or "systematic errors." They are characterized by being always in the same direction, that is, either positive or negative. Reagent blanks, incorrectly calibrated instruments, extraction efficiencies less than 100 percent and so on, are examples.

Bias errors may be constant, for example, reagent blanks in spectrophotometry, or proportional to analyte concentration, as in solvent extraction efficiencies. Fortunately, bias errors can usually be identified by proper method validation techniques, either by using standard reference materials or by comparing with methods of known accuracy. Once identified, the bias can be eliminated or applied to the results as a correction, although elimination is always to be preferred to correction as a control method.

Random Errors

Experience shows that any measurement process, especially when carried to its limit of capability (measured to the least significant digit) will yield varying results on repetition. Even as simple a task as measuring the length of a board for carpentry is subject to error, as witness the old carpenter's dictum: "measure twice, cut once." The reason for the variability is that errors occur that are not subject to conscious control by the measurer. While some of these might be considered "human errors," they are not in the same class as "blunders" since they can not be eliminated. Examples are parallax in reading calibration marks on equipment and instruments; noise in instrument response; variations in time of drying, boiling, and so on; and sample-handling errors; minor deviations from homogeneity in samples or solutions that are not well mixed.

As intimated above, random errors are only revealed on repetitive measurements, when results will be found to vary. Two characteristics distinguish random errors from blunders and bias: they are as likely to be positive as negative, and the larger the error the less frequently it occurs. In other words large errors are less likely than small.

These two characteristics imply that there is an underlying *distribution* of random errors that can be measured and evaluated. Thus, even though random errors cannot be eliminated or controlled, their effect on the result can be estimated and used to quantitate our confidence in the accuracy of the result. To do this we need to apply the techniques of mathematical statistics. An analytical system which produces results which allow us to predict the possible variability of the results is said to be in statistical control. The two major tasks of quality control are to control, monitor, and estimate variability in the analytical data, and to eliminate the effects of bias (and blunders, if possible).

STATISTICS

The science of statistics was developed to assist in the description, evaluation, and manipulation of large quantities of data. The mathematics involved are essentially simple, and rarely require more than basic arithmetic. However, the underlying principles and philosophy are often quite subtle and so the interpretation of statistical results can be difficult and involved.

In today's world we are inundated with statistics: political polls, consumer surveys, sociological studies, sports "stats," and the like are in every newspaper and TV news program. Unfortunately, these results are usually reported very uncritically, by reporters who are unqualified to judge their validity. The reporters rely on the generators of the data and these persons often have an axe to grind, a point to make, a grant to obtain, or whatever.

As an example we frequently read of surveys taken of high school students regarding drug and alcohol abuse. These are usually viewed with alarm by sober, responsible citizens as indicative of undesirable behavior on the part of the younger generation, and solemn conclusions are drawn about the increase or decrease in drug usage. I was warned many years ago about the dangers of paying too much attention to these surveys by a member of the younger generation (my own son), because, he said, many high school students think it is great fun to lie on these surveys as a kind of joke on the establishment. Anyone familiar with teenagers will recognize this as a plausible thesis. Yet these surveys go on and are taken quite seriously. This is not to say that there is not a drug problem in the country, but only that a questionnaire given to students (especially when all answers are anonymous, as they must be) will not give accurate, quantitative answers regarding its magnitude.

Another example is the telephone polls taken and reported by television stations, where viewers are asked to call in with their answers to questions. After solemnly assuring viewers that "this poll is *not* scientific" the announcer will just as solemnly give the results. If the poll is not scientific, why bother to hold it and especially why bother to report the results as if they mean something?

All this is by way of preamble to warn the reader that, while statistics may be very easy to calculate, especially with modern computers, interpretation of results and drawing conclusions may not be all that easy. Interpretation usually depends on assumptions regarding the population from which the data are drawn. For example, in our case of applying statistics to random error and evaluation of accuracy, two critical assumptions are that there is no bias in the data and no blunders. Another very important assumption is that the data being evaluated are drawn from a population which can be described by the normal distribution (cf. below).

Elementary Statistics

If N analyses are run on a given, homogeneous sample, N different results will be obtained. Some of them may coincide, but in general they will cover a range of values. (In fact, the range is a statistic defined as the difference between the largest and smallest values.) Looking at these N data, it is apparent that they tend to cluster around a value somewhere in the center of the group, and the first step in the analysis of the data is to calculate this central tendency.

This is done, of course, by calculating the ***mean*** or ***arithmetic average***, by adding all N numbers together and dividing by N.

Although the arithmetic average is by far the most common measure of the central tendency, there are other statistics that are sometimes used. In some microbiological work, for example, plate counting, the ***geometric mean*** is often used. This is calculated by multiplying all the N values together, and taking the Nth root of the product. (The easiest way to do this is to take the arithmetic mean of the logarithms of the N data and then take the antilog to obtain the geometric mean.)

Another measure of the central tendency sometimes used is the ***median***, or the value which divides the group in half, that is, half the numbers are larger than the median and half are smaller. This statistic is often used in economics and other social sciences because it may be more representative of the data than an average. For example, the median value of personal income may be more informative of a given population than the mean, since a few very large incomes may skew the data toward a mean that few members of the population possess. The author has also seen the median used in a flexural strength test of textile fibers since the median correlated better with end-use applications.

The ***mode*** is another measure of the central tendency. The mode is simply the most frequent value in the collection of data. It is possible for the data to have more than one mode, that is, two or more values with the highest frequency. In this case the data are said to be bimodal or multimodal.

Besides the central tendency of the data, a number that represents the spread or "dispersion" of the data can be calculated. There are several ways of doing this. One has already been mentioned: the ***range*** or difference between the highest and lowest numbers. The range is sometimes used in analytical chemistry because it is easy to calculate. However, it suffers from the disadvantage of being very sensitive to very high or very low values, which may be outliers (due to blunders).

Another measure of the spread of data is the ***average deviation***.

This is calculated by summing the differences of each number from the mean, using the absolute value, that is, disregarding the sign of the difference, and then dividing by N, the number of data involved. The average deviation is still used in analytical chemistry, especially when N is small, because it is easy to calculate and interpret and gives a good, intuitive feel for the spread of data. Its disadvantage is that it is not conducive to further statistical analysis or manipulation.

By far the most common measure of the dispersion of the data is the ***standard deviation***. This is calculated from the equation

$$S = \left[\frac{\sum(X_i - \overline{X})^2}{N - 1}\right]^{1/2} \tag{2-1}$$

where $\overline{X}$ is the arithmetic mean, and the X_i are the individual data in the collection. Other equations, derivable from Equation (2-1), that are sometimes used because they are easier to calculate are

$$S = \left[\frac{\sum X^2 - \left(\sum X\right)^2/N}{N - 1}\right]^{1/2} \tag{2-1a}$$

and:

$$S = \left(\frac{\sum X^2 - N\overline{X}^2}{N - 1}\right)^{1/2} \tag{2-1b}$$

The $N - 1$ factor in these equations is called the *degrees of freedom* associated with the statistic S. Since all N numbers are used in the calculation of $\overline{X}$, one degree of freedom has been used up prior to the calculation of S, that is, there are only $N - 1$ independent data left.

Another measure of dispersion which is frequently used in analytical chemistry is the ***variance***, which is simply the square of the standard deviation. The variance is useful in analyzing the cause(s) of the dispersion of the data.

The dispersion of the data is often expressed as the coefficient of variation or CV, which is the standard deviation divided by the mean and multiplied by 100, or the relative standard deviation (RSD), which is simply the standard deviation divided by the mean.

$$CV = 100\left(\frac{S}{\overline{X}}\right) \tag{2-2}$$

$$RSD = \frac{S}{\overline{X}} \tag{2-3}$$

The following table and equations show sample calculations of these various statistics, using a set of 13 analyses of a homogeneous sample of breakfast cereal for percent protein.

Sample No.	**Data** X	$X - \overline{X}$	$(X - \overline{X})^2$
1	13.4	0.05	0.002
2	13.6	0.25	0.063
3	13.4	0.05	0.002
4	12.5	−0.85	0.723
5	13.8	0.45	0.203
6	12.7	−0.65	0.423
7	12.7	−0.65	0.423
8	13.1	0.25	0.063
9	13.6	0.25	0.063
10	13.8	0.45	0.203
11	13.5	0.15	0.023
12	13.8	0.45	0.203
13	13.7	0.35	0.123
Sum:	173.6	4.85*	2.517

*4.85 is the sum of the absolute values of the deviations, i.e., the sum of the deviations regardless of sign.

$$\text{Average (arithmetic mean)}, \bar{X} = \frac{\sum X_i}{N} = \frac{173.6}{13} = 13.35$$

$$\text{Average deviation} = \frac{\sum |X - \bar{X}|}{N} = \frac{4.85}{13} = 0.373$$

$$\text{Standard deviation} = \left[\frac{\sum (X - \bar{X})^2}{N - 1}\right]^{1/2} = \left(\frac{2.517}{12}\right)^{1/2} = 0.46$$

$$\text{Variance} = (\text{standard deviation})^2 = 0.212$$

$$\text{Relative standard deviation (RSD)} = \frac{s}{\bar{X}} = \frac{0.46}{13.35} = 0.0345$$

$$\text{Coefficient of variation (CV)} = 100 \times \text{RSD} = 3.45\%$$

(It is customary in calculating these statistics to carry out calculations to one more significant figure than is in the raw data, to minimize round-off errors.)

The Normal Distribution

It should be noted that most of the previous statistics can be calculated for any set of numbers, even a set of random numbers, or the digits from zero to ten. However, such calculations are essentially meaningless. The mean, standard deviation, variance, and so on, are only meaningful when considered in conjunction with the *probability distribution* of the data set. The probability distribution is the underlying equation that describes the probability that a given number will show up in our data set.

In analytical chemistry the underlying distribution that is assumed to govern the occurrence of random errors is the *normal distribution,* also known as the Gaussian distribution, Gaussian curve, or bell-shaped curve. It should be noted that this is an *assumption,* which is seldom actually established for a given analytical method. There is very little work reported in the literature on the actual demonstration that this distribution holds for analytical work. Nevertheless, intuition and the previously stated facts that positive and negative errors seem to be equally likely, and the frequency of an error is inversely proportional to its size, indicate that, in most cases, the assumption is probably correct.

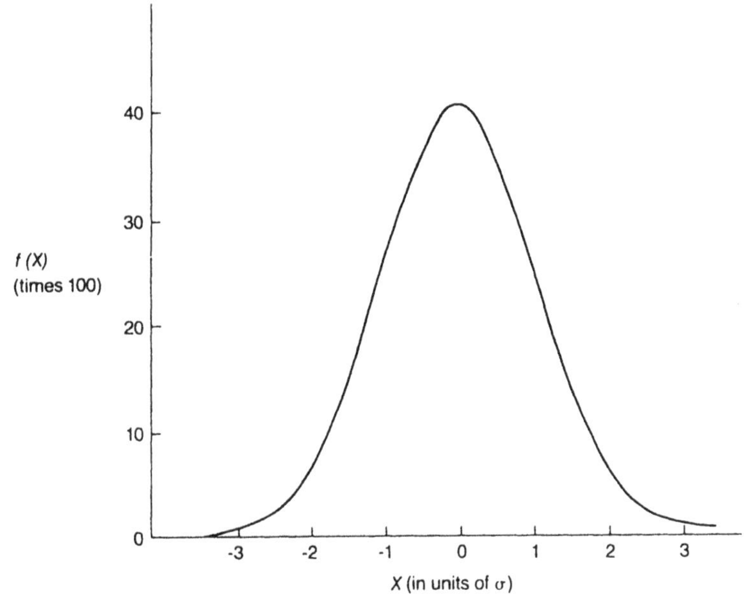

Figure 2-1. The Normal Distribution

The normal distribution can be described by an equation:

$$F(X) = \left[\frac{1}{(2\pi)^{1/2}\sigma}\right]\exp\left[-\frac{(X-\mu)^2}{2\sigma^2}\right] \tag{2-4}$$

where σ corresponds to the standard deviation and μ represents the mean. $F(X)$ is the frequency with which the value X appears. [The mean and standard deviation in Equation (2-4) are those of the "population," i.e., the infinite number of possible Xs. It is customary to use Greek letters for the statistics of the population, and Latin letters for those of a sample taken from the population.] Figure 2-1 is a plot of the normal distribution, which might underlie the data in the table on page 17. The mean was assumed to be 13.50 and the standard deviation 0.45. The X axis is calibrated in units of $\pm\sigma$.

The important point about Equation (2-4) is that it can be integrated by standard mathematical techniques, and the integral between any two points on the X axis (the area under the curve between the two points) represents the probability that a single determination by an analytical method that is governed by the curve will fall between those points. For example, the area between $+1\sigma$ and -1σ is 68.26 percent of the total area. Therefore, there is a probability of 68.26 percent that a single determination will fall in the range $\pm 1\sigma$ of the mean, or true value of the analyte. Similarly, the 2σ limits cover 95.46 percent of the area and the 3σ limits cover 99.73 percent of the area. Thus, there is a 95.46 percent probability that the determined value is within $\pm 2\sigma$ of the mean, and so on. The important fact is that the converse of these statements is also true: There is a 95.46 percent probability that the mean or true value is within $\pm 2\sigma$ of our single determination.

Thus we arrive at the important concept of the *confidence range* or *confidence limits*.

Confidence Range

The confidence range is a range of values associated with a test result within which we can state with some numerical degree of probability that the true value of the test resides.

In analytical work two confidence limits are commonly used. These are the 95 percent confidence limits, $\pm 2\sigma$, and the 99 percent confidence limits, $\pm 3\sigma$.

Practical Applications

In normal analytical work, a series of N determinations on the same sample may be run, and a standard deviation S, such as is illustrated in the table on page 17, may be calculated. This is, of course, not the standard deviation of the population, σ, but only an estimate of this statistic. The standard deviation is often called the *precision* of the analysis, but this is a misnomer, since the larger the standard deviation, the lower the precision. It is more correctly called the *imprecision* of the analysis.

Unfortunately, to calculate σ requires that we run a very large number of analyses, preferably an infinite number, which is, of course, impossible. How can confidence limits with a finite number of determinations be calculated?

This problem was solved by the chemist W. S. Gossett in the early years of this century when he published a method for calculating confidence limits for finite groups of analytical determinations. Gos-

Table
2-1. Student *t* Factor

		t Factors	
N	$N - 1$	95% Confidence Limit	99% Confidence Limit
2	1	12.71	63.66
3	2	4.30	9.93
4	3	3.18	5.84
5	4	2.78	4.60
6	5	2.57	4.03
7	6	2.45	3.71
8	7	2.36	3.50
9	8	2.31	3.36
10	9	2.26	3.25
11	10	2.23	3.17
12	11	2.20	3.10
13	12	2.18	3.06
14	13	2.16	3.01
15	14	2.15	2.98
20	19	2.09	2.86
∞	∞	1.96	2.58

sett published under the name "Student" (said to be because he was fearful of his employer learning that he was wasting his time working on statistical theory). He introduced the concept of the t factor, by which S was to be multiplied to yield a given confidence limit. Table 2-1 is a list of t factors for 95 and 99 percent confidence limits, as a function of $N - 1$, the number of degrees of freedom in the calculation. Note that the t factors are quite large for $N = 2$, but rapidly decrease as N increases, with 1.96 for the infinite case and 95 percent confidence, and 2.58 for the 99 percent confidence limits. Looking at the data in the table on page 17, we see that the standard deviation is 0.46 and $N = 13$ (12 degrees of freedom). From Table 2-1 we see that t for $N - 1 = 12$ is 2.18 for the 95 percent confidence level. Therefore, the 95 percent confidence range for any *single* determination using this method and analytical system is $\pm 2.18 \times 0.46$ or ± 1.00. Thus, if an analyst reports a result of 13.2 we can state that the true value lies between 12.2 and 14.2 with 95 percent confidence. What this means is that 95 times out of a hundred (or 19 times out of 20) when we make such a statement we will be correct.

Many analytical chemists use the 2 and 3 multipliers for 95 and 99 percent confidence limits with limited data. While this may be satisfactory as a rough approximation, it is not correct for accurate work. Unless N is at least about 50 for the 95 percent case or 15 for the 99 percent case, the confidence limits will be understated. It is best to use a table of t factors.

Standard Deviation of the Mean

It is felt that the average of N determinations is a better estimate of the true value than any single determination, and this is true. The standard deviation of the mean of N values is given by

$$S_{\overline{X}} = \frac{S}{N^{1/2}} \tag{2-5}$$

Thus, the standard deviation of a mean is reduced by the factor $1/N^{1/2}$ and hence the precision is increased by the same amount. (Statisticians, for no discernible logical reason except possibly to confuse chemists, usually call the standard deviation of the mean the standard "error" of the mean.)

Note that the square root function has important practical implications. Many analysts seem to feel that the average of duplicate results must somehow be twice as good as a single determination. Actually, it is only $\frac{1}{2}^{1/2}$ or 0.707 times as good, an improvement of only 30 percent. To obtain a result that is twice as good, four determinations are necessary, and nine for a threefold improvement. It should be apparent that the work required for a marginal improvement in precision soon becomes impractical, and a better method with a smaller standard deviation should be sought, rather than continuing to run multiple samples. (For example, to improve precision by a factor of ten, 100 replicates need to be run and averaged.)

For the data in the table on page 17 the mean is 13.35, the standard deviation is 0.46, and N is 13. The standard deviation of the mean then is $0.46/(13)^{1/2}$ or 0.13, and the 95 percent confidence limit is 0.13 × 2.179 or ±0.28, a considerable improvement over the value of ±1.00 for a single determination, but at the cost of 13 times as much work.

The Variance

The variance is not often reported in analytical work because the units of the variance are the square of the units of the mean but, as mentioned previously, it is often used in analyzing the causes of the dispersion of the data. In fact mathematicians have developed a whole branch of statistics called the analysis of variance (ANOVA).

One useful property of the variance is that it is summative, that is, if the total variance of a set of data is due to two or more sources of random variation, which are independent of each other, the total variance is the sum of the variances due to each cause. For example, a simple chemical analysis may require weighing a sample followed by a titration. Random errors due to weighing are independent of random errors due to titration, so the total variance is given by

$$V_t = V_g + V_l \tag{2-6}$$

where V_g is the gravimetric variance and V_l is the titration variance.

Another example would be in the analysis of a series of samples from a bulk material exhibiting nonhomogeneity. In this case if we know the variance due to the analysis (possibly from results on homogeneous samples) we can calculate the inhomogeneity of the material by

$$V_h = V_t - V_m \tag{2-7}$$

where V_t is the total variance and V_m is the variance due to the measurement.

The F Test

A frequent problem in analytical work involves determining whether two sets of data have significantly different variances (or standard deviations). The two sets of data may be from different analytical methods, different analysts, or different samples, for example. The problem to solve is whether the methods, analysts, or samples are significantly different from each other.

Mathematicians have developed a simple test for evaluating whether the variances are significantly different to any degree of confidence. The F value is calculated by dividing the larger variance by the smaller:

$$F = \frac{V_1}{V_2} \tag{2-8}$$

where $V_1 > V_2$.

The calculated F value is compared with a table of F values. Each table is entered with the two variables: $N_1 - 1$ and $N_2 - 1$, the degrees of freedom associated with the two variances. If the calculated F value is greater than the value in the table, then the difference in variances is significant to the confidence level given for the table. The appendix contains tables for the 95 percent and 99 percent confidence levels. Other tables are available in standard statistics texts or compendia of statistical tables. The F test can be a valuable tool for comparing precision of analytical methods, skill of analysts, homogeneity of samples, and other similar applications.

Pooling Variances

Occasionally it may happen that a series of samples of different analyte concentrations may be run in replicate and a series of standard deviations calculated, one for each sample. It may be desirable to calculate an overall estimate of imprecision by pooling the individual variances. If there are k samples, and N_1 is the number of analyses run on the first sample, N_2 the number on the second sample, and so on, the variances may be pooled using the equation

$$V = \frac{\sum V_i(N_i - 1)}{\sum N_i - k} \tag{2-9}$$

where N_i is the number of replicates run on the ith sample.

The most common application of this equation is in the case of duplicate analyses. Many laboratories run samples in duplicate as a quality control measure. In pooling variances from duplicates, Equation (2-9) reduces to

$$V = \frac{\sum R_i^2}{2N} \qquad (2\text{-}10)$$

where R_i is the *range* of the *i*th sample, that is, the difference between the duplicate values, and N is the total number of samples run (not the number of analyses, which is $2N$). The standard deviation calculated from Equation (2-10) by extracting the square root of V applies to a single determination by the method, not the mean of the duplicates.

Equation (2-10) is a useful tool for evaluating the imprecision of a method, and can also be used to compare methods, analysts, and so on, by use of the F test to compare variances. The number of degrees of freedom associated with V in Equation (2-10) is N, the number of samples run. Table 2-2 shows an example of calculations for the standard deviation using 26 samples that were run in duplicate. The analysis was for percent moisture in animal feeds of different kinds, a simple drying test. As is well known there is no way of checking the accuracy of such a test, since drying may remove volatile materials other than moisture and there are no adequate standard materials with known moisture contents. Nevertheless we can calculate a standard deviation for the method, and hence a confidence interval for any single determination, under the assumption that the test is accurate.

Equations (2-9) and (2-10) will only give meaningful results if the variance (and standard deviation) are independent of the analyte concentration, a condition that is not always true. If the variance increases with the analyte concentration, which is often the case, the estimated imprecision will be too large for low analyte concentrations and too small for high concentrations. To determine if this is the case, the variance for high concentration levels can be compared to that at low levels, using the F test. The data in Table 2-2 were so tested and the variance between high values of moisture and low values was found not to be significantly different. Note the wide variation in moisture levels in these samples from about 6 percent to 45 percent.

Table
2-2. Standard Deviation from Duplicate Analyses

Sample No.	Analysis 1	Analysis 2	Range	Range^2
1	46.70	46.90	0.20	0.0400
2	7.10	7.10	0.00	0.0000
3	6.77	6.71	0.06	0.0036
4	9.06	9.39	0.33	0.1089
5	11.82	11.72	0.10	0.0100
6	9.37	9.24	0.13	0.0169
7	9.88	9.94	0.06	0.0036
8	8.21	8.13	0.08	0.0064
9	10.48	10.45	0.03	0.0009
10	9.86	9.73	0.13	0.0169
11	7.52	7.44	0.08	0.0064
12	45.86	46.26	0.40	0.1600
13	11.61	11.37	0.24	0.0576
14	10.10	9.97	0.13	0.0169
15	9.72	9.47	0.25	0.0625
16	9.56	9.42	0.14	0.0196
17	10.54	10.47	0.07	0.0049
18	11.27	11.28	0.01	0.0001
19	10.57	10.43	0.14	0.0196
20	7.30	7.49	0.19	0.0361
21	10.19	9.70	0.49	0.2401
22	10.31	10.26	0.05	0.0025
23	8.46	8.14	0.32	0.1024
24	12.15	12.28	0.13	0.0169
25	8.00	7.70	0.30	0.0900
26	7.71	7.76	0.05	0.0025
Sum			4.11	1.0453

$$\text{Mean range} = \frac{4.11}{26} = 0.158.$$

$$\text{Standard deviation} = \left[\frac{1.1542}{(2 \times 26)}\right]^{1/2} = 0.149.$$

Comparison of Means

It frequently happens in analytical work that we would like to compare the means of two different sets of data to determine if the difference is significant or simply due to random variation. The data sets may represent, for example, different analytical methods, different analysts, or a different time sequence on a control chart (cf. Chapter 3). Two techniques are used, depending on whether it can be assumed that the variances of the two sets of data are the same or different. In the first case, where the data may be assumed to have the same population variance, first calculate the pooled variance for the difference between the two means, using Equation (2-9) for $i = 2$:

$$V = \frac{V_1(N_1 - 1) + V_2(N_2 - 1)}{N_1 + N_2 - 2}$$
$$S = V^{1/2} \tag{2-11}$$

Using this value of S calculate the standard deviation for the difference between the two means from

$$S_d = S\left(\frac{N_1 + N_2}{N_1N_2}\right)^{1/2} \tag{2-12}$$

Then calculate the 95 or 99 percent confidence limits for d from S_d and the t value for $N_1 + N_2 - 2$ degrees of freedom.

Example 1: Two analysts run 10 determinations each on a check standard used for a control chart. Analyst 1 obtains a mean result of 1.96 with a standard deviation of 0.17. Analyst 2 obtains a mean of 2.02 with a standard deviation of 0.19. Is there a significant difference between the two means?

The variance calculated from Equation (2-11) is

$$V = \frac{9(0.17)^2 + 9(0.19)^2}{18} = 0.0325$$

or

$$S = (0.0325)^{1/2} = 0.18$$

The difference between the means is 2.02 − 1.96, or 0.06. The standard deviation for this difference is calculated from Equation (2-12):

$$S_d = 0.18\left(\frac{20}{100}\right)^{1/2} = 0.079$$

The t factor for 18 degrees of freedom at the 95 percent confidence level is 2.09. Therefore, the 95 percent confidence range for d is 0.06 ± 2.09 × 0.079 or 0.06 ± 0.16, that is, −0.10 to +0.22. Since zero is within this range we can conclude that there is no significant difference between the two means.

The second case occurs when the two sets of data do not arise from the same underlying distribution, or at least there is reason to suspect that this is the case. A good example would be in comparing two analytical methods, both run in replicate on the same sample. In this case the variance of the difference in means is calculated from Equation (2-6):

$$V_d = V_1 + V_2$$

where V_1 is given by S_1^2/N_1 and a corresponding value is calculated for V_2. Then S_d is simply calculated from the square root of V_d. Using the t factor for $N_1 + N_2$ degrees of freedom, the 95 percent confidence limits for d may be calculated as in the previous case.

Example 2: A laboratory is considering modifying an analytical method to decrease the time of the analysis. A suitable standard material is analyzed five times by the old method and the new, using the same analyst, instrument, reagents, and so on. Because the modification is a major one, it cannot be assumed that the imprecision (standard deviation) is the same for the two methods.

The original method yields a mean value of 4.52 percent for the five analyses with a standard deviation of 0.15 percent. The modified method yields 4.32 percent for the mean, with a standard deviation of 0.22 percent. Are the two methods identical with regard to precision and accuracy? Or, in other words, are the means and standard deviations significantly different, at some confidence level?

Calculations of F yield

$$F = \frac{(0.22)^2}{(0.15)^2} = 2.15$$

The critical value of F for four degrees of freedom in the numerator and denominator, at the 95 percent confidence level, is 9.60. Therefore, it cannot be concluded that the variances are significantly different without further evaluation. Calculate the variances for the two methods from the standard deviation of the means, that is, S^2/N:

$$V_1 = \frac{(0.15)^2}{5} = 0.0045 \qquad V_2 = \frac{(0.22)^2}{5} = 0.00968$$

and the variance of the difference between the means by adding them:

$$V_d = 0.0045 + 0.00968 = 0.01418$$

$$S_d = (0.01418)^{1/2} = 0.119$$

The t factor for the 95 percent confidence level with 10 degrees of freedom is 2.228. Therefore,

$$95\%\text{CL} = \pm 2.228 \times 0.119 = \pm 0.265$$

$$d = 4.52 - 4.32 = 0.20$$

Thus,

$$d = 0.20 \pm 0.265$$

or d may range from -0.065 to 0.465. Since zero is within this range we cannot conclude that the two means are significantly different at the 95 percent level of confidence.

These are only some of the more practical examples of the use of statistics in the analytical chemistry laboratory. Others will be covered in later chapters of this book, especially in sections dealing with quality control techniques and analytical method validation. Every analytical chemist should have some training in the use of statistical methods of data evaluation. Statistics is not only a useful tool, but it raises the consciousness of analysts regarding the possibility of error and the magnitude of errors in their reported data.

CHAPTER

3

Statistical Quality Control Techniques

In Chapter 1 the concept of the analytical system was introduced. To reiterate, the analytical system consists of all elements in the laboratory that can impact on the result of an analysis. This includes the laboratory's physical facilities, personnel, analytical methods, operational procedures, and instruments and other auxiliary equipment. To control quality it is necessary to control the elements that make up the analytical system.

However, basic to the idea of controlling a system is the need to monitor its performance. Thus, much of the content of quality control is the techniques and methods used to monitor the system in its day-to-day operations. This chapter will discuss some of the more common techniques used in today's laboratories.

CONTROL CHARTING

The idea of control charts for analytical system monitoring was borrowed from manufacturing technology where the concept has been in use since the 1930s. Shewhart[1] developed the idea of periodically sampling the production line and testing the samples for quality characteristics. A graph was then constructed with a center line corresponding to the average value of the characteristic and two lines showing the 99 percent confidence limits. The graph was continuously updated and evaluated for evidence that the system was in statistical control. A point outside the 99 percent limit line was evidence of loss of control, and a series of points on one side or the other of the center line was evidence of a trend or nonrandom variation in the system.

Workers at the National Bureau of Standards appear to have been the first to apply Shewhart control charting to measurement processes.[2] There is an interesting difference between the production control chart and the measurement chart. In the production case, the measurement is considered to be without error (or with minimum error, since averages of several tests are often used) and all variation is assigned to the sample as indicative of the variations in the production process. In the measurement case the sample characteristic is assumed to be invariant and variation is due to the testing system.

In modern analytical applications of Shewhart-type charts, lines corresponding to the 95 percent confidence limits are also used and are called the *warning limits,* while the 99 percent limits are called the *control limits.* A point between the 95 percent limits and the 99 percent limits is considered evidence of possible problems with the analytical system since this should occur only 5 or less times in 100 tests, or once in every 20 on average. On the other hand, a point outside the 99 percent limits is presumptive evidence of loss of control since this should happen only once in about 300 tests on the average (technically 3 times in 1000), if only random variation is occurring.

The use of control charts in the analytical laboratory today is generally recognized as the most valuable technique for continuously monitoring the analytical system. The control chart not only indicates serious trouble with the system, but can indicate trends toward trouble before they become problems. Three types of control charts are in wide use in laboratories: the *X*-chart, which is similar to the basic Shewhart chart; the spiked-sample chart, which is also similar; and the range or *R*-chart, which is based on the difference between duplicate analyses. These three types of charts are described below.

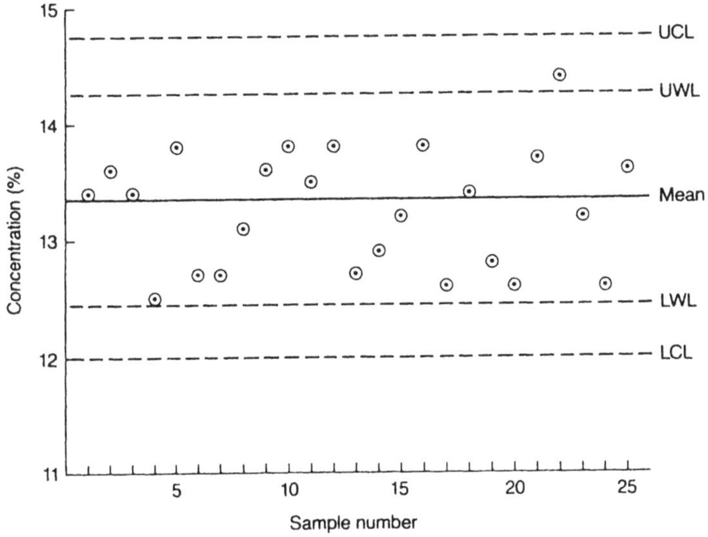

Figure 3-1. Example of *X*-control chart.

X-Charts

The basic idea of the *X*-chart is the same as the Shewhart chart with the difference in interpretation noted above. The technique can be applied to any standard repetitive analysis that is run on a series of unknown samples with the same or similar substrates. A standard reference material or "check standard" is selected that is analyzed along with the unknown samples. One analysis of the check standard is run with each batch of unknowns, or, if a large number of unknowns is run in a batch, one check standard for each 10 or 20 unknowns.

After 15 to 20 check standard analyses the mean and standard deviation of the data are calculated and a control chart constructed. Figure 3-1 is an example of such a chart. The center line represents the mean, the two outer lines represent the upper (UCL) and lower (LCL) control limits, or 99 percent confidence limits, and the two lines closest to the mean line are the 95 percent confidence limits, or upper (UWL) and lower (LWL) warning limits.

Once the chart is constructed it is continued indefinitely, plotting new points as they are generated as part of the analytical process and continuously recalculating the mean and standard deviation. As time goes on a considerable body of data is built up and very accurate estimates of the mean and standard deviation result. After the first 30 or so points the confidence limits are usually assumed to be twice and three times the standard deviation. While not completely accurate, this is considered acceptable.

If only one check standard analysis is run each day, about 260 data points will accumulate in the course of a year's operation. Manual maintenance of the control chart will be cumbersome and time-consuming. It is more practical to keep a chart for each 25 or so analyses. In fact, the use of a computer greatly simplifies the use of control charts, especially if a large number of charts are kept. Data files can be used to accumulate data, statistical programs are available to calculate mean and confidence limits, and programs to display the chart of the last 25 or so points are available. "Hard copy" printouts of the charts can be made periodically, every week or month, for scrutiny by the quality assurance personnel or laboratory supervision. Variation of the weekly or monthly mean or standard deviation from the previously established value can be evaluated for significance by the F test for the standard deviation or the technique given in Chapter 2 for comparing means. With a computerized system a decision-making routine can be used to flag a result that exceeds the 95 percent confidence limits when data are entered, so that immediate action can be taken.

Some analysts may question the need for charting at all, on the basis that they can look at the data and extract the same information. This is not generally true. The human eye and brain are uniquely suited to recognize patterns in graphically presented information, and this is why the visual chart is valuable. In Figure 3-1 it can immediately be seen that one point is outside of the 95 percent confidence limits, sample 22. However, since this predictably will occur 5 times out of 100, or 1 time in 20, due to random variation, this is not cause for alarm, especially since the next point is within the limits.

Figure 3-2 is another example of an X-chart. The reference standard in this case is a solution of cadmium ion, prepared at a nominal concentration of 1.0 mg/L. The mean, calculated from 188 replicate analyses, is 0.989 mg/L and the standard deviation is ± 0.0225 mg/L.

Note that the last seven points, even though within the warning limits, are all on the low side of the mean line. There is a 50 percent

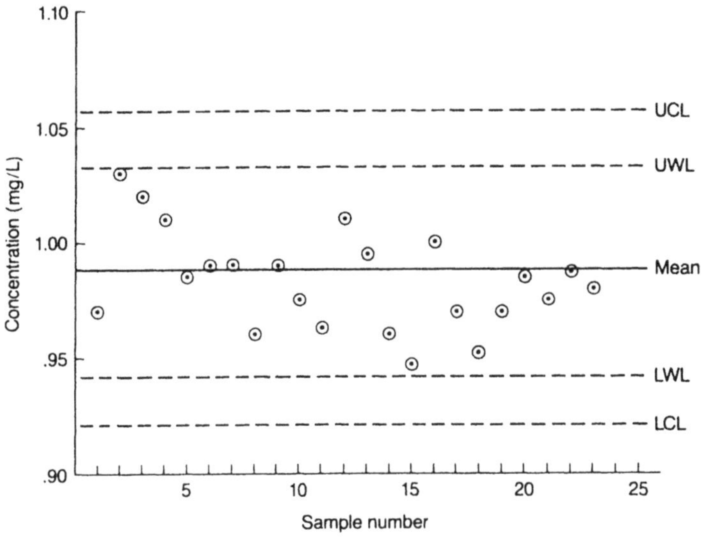

Figure 3-2. *X*-control chart showing drift.

probability or 1 in 2 chances of any given point being on one or the other side of this line. The probability of two consecutive points being on the same side of the line is 1 in 4 or 25 percent. For *n* consecutive points the probability is $1/2^n$. Therefore, the probability of the sequence of seven consecutive results on one side of the line in Figure 3-2 occurring purely by random fluctuation is $1/2^7$ or 1 in 128, somewhat less than 1 percent. This is sufficient to indicate that something has changed in the analytical system and an investigation should be made to determine the problem. Note that the error in the seven analyses is not large and the samples run during this time may or may not need to be rerun, depending on the use to which the data is put. (Note also the general downward trend as the sample number increases, another indication that something is wrong.) This is the kind of analytical error that might never be discovered without the use of a control chart, until the problem becomes much worse. Thus the regular scrutiny of control charts will alert the laboratory supervisor or Q.A. officer to trouble before it becomes a serious problem.

One analysis outside the 95 percent limits, but within the 99 percent limits, is not cause for alarm, since this can be expected to occur approximately once in every 20 analyses. In fact, if this does not occur, on average, the program for calculating the limits should be reevaluated. However, the probability of two consecutive analyses falling on one side of the mean line between the 95 and 99 percent limits is $\frac{1}{20}$ times $\frac{1}{20}$ or $\frac{1}{400}$, which would certainly be cause for an investigation, as would be only one analysis outside the 99 percent limits.

The control chart technique also gives an excellent estimate of the precision of the analytical system when a large number of replicate analyses are run, as in Figure 3-2. The difference between the mean of 0.989 and the "makeup" concentration of 1.00 mg/L is also an excellent estimate of the accuracy of the system. It should be pointed out, though, that the use of a check standard consisting of a solution of one analyte in distilled water, as was done in this case, does not address the problems of matrix effects or mutual interference, which may be present in real world samples.

The choice of a reference material or check standard is very important in control charting. The following are some characteristics the reference material should have:

1. The concentration of the analyte should be known, but this is not an absolute requirement. When dealing with analyses of natural materials such as food, soil, or minerals it is often not possible to find or prepare a standard of known concentration. An example would be protein in food by the Kjeldahl method. A good control chart system can still be set up, using a material of initially unknown concentration.

   If the concentration of the analyte in the reference material is not accurately known, an estimate of the accuracy of the analytical system is not possible. However, the control chart can still be used to estimate precision and to indicate unsuspected changes in accuracy and precision as discussed above.

   Check standards for analyses of liquids are usually easy to prepare with known concentrations of soluble analytes. However, in the case of solids, considerations of homogeneity enter the question. They must be thoroughly mixed to ensure no variability due to concentration variation.

2. The check standard material should be available in large quantities, since it is undesirable to have to change to a new batch of standard when the initial material runs out.

3. The check standard material should be stable with regard to the analyte, that is, the analyte concentration should not change with time. In some cases unstable analytes can be used if the material is kept at low temperatures, or other preservative techniques are used, such as keeping the material under an inert atmosphere, in opaque containers, or in tightly closed containers.

4. The reference material should be of a substrate or matrix the same as, or very similar to, the unknown samples being analyzed by the same system. This will tend to minimize matrix effects in evaluating accuracy and precision in the unknowns.

5. The analyte concentration in the check standard material should be in the same concentration range as is found in the unknown samples, which are analyzed at the same time, otherwise the estimates of accuracy and precision calculated from the control chart may not be applicable to the unknowns.

It is important that the check sample be subjected to *all* steps of the analytical procedure, exactly the same as the unknown samples, and that the check sample be run at the same time and by the same person as the unknown samples. Sample preparation steps and analyst skill are important parts of the analytical system. Only in this way can the control chart be said to be monitoring the system.

The frequency with which the check standard is run is a matter requiring professional judgment. It will depend on how frequently the analysis is run on unknowns, the complexity or ruggedness of the method, the cost of the analysis, and past experience. The rule of thumb most commonly used is one check with each batch of samples or one with each 10 samples if batches of samples exceed 10. The cost of running the check samples will increase with decreasing size of the sample batch. If large batches of samples are run, the above rule will increase the cost by 5 to 10 percent. If only one unknown is run on an infrequent basis, the cost of the check sample program will be 100 percent, that is, twice the cost of the analysis without the check sample program. In the latter case it may be cost-effective to save up the unknowns until a sufficient number have accumulated, provided the samples and the analyte are stable and the user's needs are satisfied.

One further point regarding control charts should be mentioned. You will note that lines are not drawn between the points on the charts in Figures 3-1 and 3-2. While this is a common practice often found in the literature, there is no real reason for doing this. In fact, the lines imply that one is plotting a continuous function, which is not the case. There is no reason to believe that if a sample were run between any two points on the graph the result would lie between the two points.

Spiked-Sample Control Charts

Spiked-sample control charts are frequently used in cases where check standards of appropriate analyte concentrations are not easily prepared or obtained. Environmental analyses of very low level analyte concentrations are a prime example. The spiked-sample chart is superficially similar to the X-chart, but instead of using a check standard, one of the unknown samples is analyzed and then spiked with a known amount of the analyte of interest. The percent recovery of the spike is calculated and plotted on the control chart. As is well known, percent recoveries in trace organic analyses can vary considerably from the theoretical 100 percent.

The control chart lines on the spiked-sample chart correspond to the mean recovery and the 95 and 99 percent limits calculated from the standard deviation of the recovery data. For analytical system monitoring, the resultant chart can be used and interpreted in the same way as the X-chart.

However, care must be taken in using the data to estimate accuracy and precision of the analytical method. The reason is that in trace level analyses it is well known that the matrix or substrate can have a profound effect on the efficiency of recovery. Therefore, unless the analyses are all run on the same or a similar matrix, the variability in the recovery will be overestimated. Provost and Elder[3] have shown that the variability in percent recovery is a function of the ratio of spike to unknown concentration, with a 95 percent confidence range of 55 to 145 percent in the common situation where the spike is equal to the background concentration and the coefficient of variation is 10 percent, for example.

With regard to the standard deviation, it should be recognized that the measurement of the spike level is the result of two determinations. Therefore, if S is the standard deviation of a single measurement, the standard deviation of the spike measurement is, at best, $2^{1/2}S$, or $1.414S$ [as in Equation (2-6), for example]. In addition, in most environmental analyses, the standard deviation is proportional to the concentration (i.e., the coefficient of variation is constant), so that the precision measured at the spike level will be lower than that at the analyte level.

The mean recovery level, calculated from the spiked-sample chart, is sometimes used to correct the results obtained for the unknown samples. This is probably not a good practice for the reasons given above, especially if there is no guarantee that the matrix effects are the same for the spike and the unknown.

Nevertheless, as mentioned previously, the spiked-sample chart may be used to monitor the analytical system to detect any trends or out-of-control situations. It is important to be sure that the spike is applied to the original sample, and not to a derived solution, so that the entire analytical procedure is monitored, including any sample preparation steps.

R-Charts

It is common practice in many analytical laboratories to run duplicate analyses at frequent intervals as a means of monitoring the imprecision of the analysis and detecting out-of-control situations. This is often done for analyses for which there are no suitable check standards available, such as suspended solids in wastewater or moisture determinations in a variety of materials. Usually the mean of the duplicates is reported and the difference between the duplicates, or range, is examined for acceptability. Frequently there is no quantitative criterion for acceptability. The use of the duplicate range in a control chart is one system of deciding acceptability of the individual ranges.

Youden and Steiner[4] have published a distribution table for the variation of duplicate differences or ranges. This is shown in Table 3-1. Based on this table it can be shown that 50 percent of the ranges are below the $0.845\bar{R}$ value, 95 percent are below $2.456\bar{R}$ and 99 percent are below $3.27\bar{R}$, where $\bar{R}$ is the average range for a set of duplicate analyses.

Using these factors a control chart can be set up for the differences between duplicate analyses, or ranges of the duplicates. After 15 or 20 duplicates are run, the average range is calculated, and the mean and control limit lines are drawn, using the 95 and 99 percent

Table
3-1. Distribution of Duplicate Ranges

Difference (Range)	Percent
Less than $\overline{R}$*	57.5
Between $\overline{R}$ and $1.5\overline{R}$	19.4
Between $1.5\overline{R}$ and $2.0\overline{R}$	12.1
Between $2.0\overline{R}$ and $2.5\overline{R}$	6.4
Greater than $2.5\overline{R}$	4.6
	100.0

*$\overline{R}$ is the mean value of R.

Source: Reprinted by permission of the Association of Official Analytical Chemists.

factors described above. The 50 percent line is also drawn on the chart, using the $0.845\overline{R}$ factor. The resulting chart is then used to continually monitor the imprecision of the analysis.

Figure 3-3 is a duplicate control chart of the data in Table 2-2, based on duplicate analyses for moisture in animal feeds. The solid line on the chart represents the 50 percent line and the two dashed

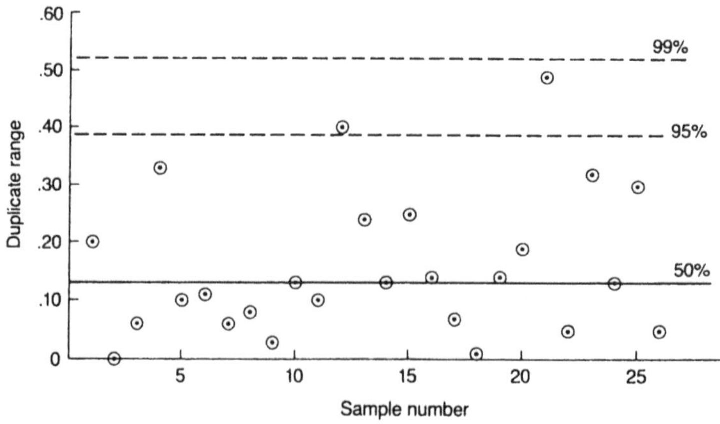

Figure 3-3. Example of R-control chart.

lines are the 95 and 99 percent confidence limit lines. Note that the chart is visually quite different from the X-charts previously considered. The distribution is quite asymmetric, with most of the points being jammed below the 50 percent line. This is because the range cannot be lower than zero since it is the absolute difference between two positive numbers.

Theory states that 50 percent of the points will be above this line, and 50 percent below. By actual count in Figure 3-3 there are 12 below, 9 above, and 5 points on the line. Note that sample 12 shows a range just outside the 95 percent confidence line. This may be expected in a set of 25 samples. Sample 21 is just within the 99 percent limit and would be cause for further investigation, except that further analyses returned to the normal range.

In interpreting duplicate control charts the 50 percent line plays the same role as the mean or average line in X-charts. If more than five or six points in succession fall on one side or the other of the line, it is a strong indication that something has changed and should be investigated. If the points are on the high side, the precision has deteriorated, while if they are on the low side, it has improved.

In setting up and interpreting duplicate control charts adhere to the same precaution discussed in Chapter 2 under statistical analysis of duplicate data. The standard deviation should be known to be independent of the analyte concentration or demonstrated to be independent. If it is not, it may be necessary to set up separate charts for different ranges of analyte concentration. In some laboratories, the percent range or range divided by the mean of the duplicates is plotted. While this seems to give acceptable results, at least in some cases, to the best of the author's knowledge, there is no theoretical basis for this type of chart.

Duplicate range control charts should be maintained continuously in the same manner as X-charts. Computers are a great help. Usually only the last 20 to 25 points need be displayed on a periodic basis for examination by the laboratory supervisor or quality assurance officer. The computer must be programmed to calculate the cumulative sum of the squares of the ranges, so that the variance and standard deviation can be easily calculated and displayed [cf. Equation (2-10)]. The significance of any apparent change in the precision, as shown on the chart, can be evaluated by the F test [Equation (2-8)] applied to the most recent value of the variance compared to the overall value of the variance.

In summary, duplicate control charts are extremely useful for monitoring the precision of analyses for which there are no acceptable reference check materials. Since they are based on the difference between two results, errors due to bias are effectively canceled, and no conclusions can be drawn regarding the accuracy of the analysis.

BLIND SAMPLES

The term *blind sample* has unfortunately received two different interpretations in recent years. Originally, the term was restricted to a sample of known concentration that was submitted as a run-of-the-mill unknown, with the analyst unaware that there was anything special about the sample. Only the laboratory supervisor or quality assurance officer knew that it was a check sample. Recently the term has been used by some analysts to designate merely a sample of unknown concentration submitted as a check sample, with the analyst fully aware of the sample's origin, with the term *double-blind sample* used for the original type of sample.

In this book the original definition of the term is retained, reserving the expression "performance evaluation sample" for the second type (remembering that there are other types of performance evaluation samples).

The major objective of a quality assurance/quality control program is to determine the quality, that is, accuracy and precision, of the data generated by the laboratory in its daily operation on run-of-the-mill samples being submitted on a more or less regular basis. The best way to determine this is with a blind-sample program. The advantage of the blind-sample program is that the standard sample is treated exactly as though it were an ordinary sample. Running check standards and other performance evaluation samples will not achieve this since the analyst knows that the sample is special and consciously or unconsciously will tend to be more careful with the sample.

One indication of this is that blunders or human errors of large magnitude will seldom occur in control charting where check standards are run or in performance evaluation samples, while they will be found in a blind-sample program. In fact a blind-sample program is one of the best ways of estimating the frequency of occurrence of blunders. It may also indicate which members of the laboratory staff are more prone to such errors, so that corrective action may be taken. A blind-sample program may also be used to evaluate analysts, using the statistical techniques given in Chapter 2.

Some laboratory managers and supervisors may feel uncomfortable in establishing and maintaining a blind-sample program in the mistaken belief that somehow this is not playing fair with the analysts and may engender a feeling that management does not trust the employees. With regard to fairness, it is certainly a manager's right and even duty to evaluate employees and the whole analytical system, and the blind-sample program is probably the best way of doing this. Employees who are fully educated in the purpose and objectives of the quality assurance program will recognize the value of the program and will have no reason to object. In fact the knowledge that a blind-sample program is in place may have a salutary effect on performance. Managers should be wary of using a blind-sample program as a punitive measure, for example, calling an analyst in for an explanation every time a 95 percent confidence limit is exceeded, for this will only cause resentment of the program. It is best to keep the results strictly confidential unless there is good reason for making them known.

In principle, setting up a blind-sample program is essentially similar to a control chart program, except that the blind samples are submitted by the quality assurance officer, and results are reported to him or her. Either X-charting or R-charting may be used. Choice of check standard should follow the same principles as given above for the X-chart program. Duplicate blinds may be submitted by splitting an unknown and assigning different sample numbers to the two samples or by resubmitting a sample after analysis under a new sample number.

Although the principle behind the blind-sample program is simple, there are practical difficulties in setting it up. It is necessary in most cases to have the cooperation of the laboratory supervisor, which may put him or her in a situation involving a slight conflict of interest. One of the biggest problems is the use of a dummy client name, which is necessary in most laboratories since this is information which accompanies the sample.

If a real client's name is used, there is the danger that results will be reported to him or her, if the blind-sample result is not intercepted by the quality assurance officer or the laboratory supervisor. This can lead to confusion. On the other hand, if a fictitious client name is used, there is the danger that the analyst may discover that the client is nonexistent, with further confusion. The best system is to seek a cooperative client who will permit the use of his or her name and will return results to the Q.A. officer as received.

A blind-sample program can be set up for any analysis, but because of the complexity of the scheme it is better suited to the more important analyses, either those run on a large volume of samples or those whose results are most important to the clients. The frequency may be much less than that used with control chart check standards. If one blind sample is submitted every 2 weeks, by the end of the year 26 samples will have been run, which will provide a good database for evaluating accuracy and precision.

PERFORMANCE EVALUATION SAMPLES

In one sense, all the samples described in this chapter—control chart check standards, duplicates, spiked samples, and blind samples—may be designated as *performance evaluation* (PE) samples since they are run for the purpose of evaluating laboratory performance. However, the term is generally used to designate samples submitted by an organization external to the laboratory to evaluate performance.

There are three types of PE samples. The first comprises those submitted in support of an accreditation program, for example, the EPA program for accrediting laboratories for drinking water testing, or the National Institute of Occupational Safety and Health PAT (Performance Analytical Test) samples for accreditation by the American Industrial Hygiene Association.

In these types of PE samples, the concentration of analytes is known to the accrediting body, but not to the laboratory. The laboratory is judged on the basis of whether results reported are acceptable or not. Acceptability is based on the previously established 95 and 99 percent confidence limits for the analysis. Usually a result between the two limits will result in provisional accreditation, pending analysis of a second sample. Results outside the 99 percent limits or two consecutive results exceeding the 95 percent limits (probability $\frac{1}{20}$ times $\frac{1}{20}$ or $\frac{1}{400}$) will result in loss of accreditation until the problem is corrected.

Since achieving or maintaining accreditation can be very important to a laboratory from a professional, economic, or legal standpoint, there is great pressure on the laboratory staff to obtain acceptable results on these samples. This may lead to multiple analyses if sufficient sample is available, calibration of instruments before and after analysis, preparation of fresh reagents, use of fresh standard solutions, and assigning of the more experienced analyst to the task. The results reported are thus not indicative of the daily quality of the laboratory's work, but rather reflect the best that can be done.

This means that laboratory management and supervision should not rely on PE sample results as being representative of laboratory quality, nor should the knowledgeable client accept such data as representative. If all samples were analyzed with the same care as PE samples this would be true, but the cost of analysis for routine samples would become prohibitive.

The second type of PE samples are those received from professional organizations, which are submitted periodically to participating laboratories. The sole purpose of these programs is to provide the laboratory with a means of evaluating its performance in comparison with other laboratories running the same analyses. Examples of these types of programs are the check feed sample program of the American Association of Feed Control Operators (AAFCO) for animal feeds, and the Smalley check sample program of the American Oil Chemists Society (AOCS).

Participation in these programs is strictly voluntary, and after the analytical results are submitted to the organization, the laboratory receives a statistical analysis of all results obtained, suitably coded so that laboratory anonymity is maintained. In this way the laboratory can compare its performance to other laboratories running the same tests. Since results of these programs are strictly confidential, there is no pressure on the analysts to produce acceptable results, and these samples are excellent candidates for a blind-sample program.

The participating laboratories are frequently ranked according to their closeness to the true value of the analyte, or to the average obtained by all laboratories, excluding outliers. Such rankings should be viewed with caution, because often there is little difference between the highest ranking laboratories. However, if a laboratory consistently achieves a high or low ranking it may be indicative of the laboratory's overall quality.

The third type of PE samples are blind check samples or duplicates submitted by the laboratory's clients for the purpose of evaluating laboratory performance. It may happen that the client may complain about laboratory results, based on such samples. If this happens, management or laboratory supervision should carefully examine the basis of the complaint, before taking any action. It may be that the client's "true value" is based on shaky evidence and may not be correct. If the client is comparing two laboratories in this manner, management or supervision should investigate to be certain that the two laboratories are using the same analytical method, same type of instrument, and so on, before concluding that there is something wrong.

REFERENCES

1. Shewhart, W. A. *Economic Control of Manufactured Products*, New York: Van Nostrand Reinhold, 1931.
2. Prontius, P. E., and J. M. Cameron. *Realistic Uncertainties and the Mass Measurement Process*, NBS Monograph 103, Gaithersburg, Md.: National Bureau of Standards, 1967.
3. Provost, L. P., and R. S. Elder. Interpretation of Percent Recovery Data, *American Laboratory*, pp. 57–63 (December 1983).
4. Youden, W. J., and E. H. Steiner. *Statistical Manual of the Association of Official Analytical Chemists*, Washington, D.C.: Association of Official Analytical Chemists, 1975.

CHAPTER 4

Sampling and Quality Assurance

The subject of sampling for chemical analysis is a very wide-ranging one and worthy of an entire book on its own. To treat the subject in depth would be beyond the scope of this book. However, this chapter is concerned with those aspects of sampling that impact on the quality and quality assurance of the analytical results.

Some chemists may feel that sampling is not a proper topic for a book on laboratory quality assurance since only the quality or accuracy of the data generated on the samples as received should be of concern. However, one of the characteristics of information as a resource is that it must meet the purchaser's needs. It is a truism in analytical chemistry that, if the sample is not taken properly, there is no point in running the analysis. This is perhaps an extension of the definition of quality, but it is a very real concern to the professional analytical chemist.

The degree of involvement of laboratory personnel in sampling varies depending on the particular laboratory and its mission. In some laboratories all sampling is done by personnel outside the lab, while in others the sampling function is assigned to laboratory personnel. Regardless of who does the sampling, however, it should be a matter of concern of the laboratory manager and supervisor, since improper sampling renders the work of the laboratory useless. Since improper sampling leads to incorrect analyses, the tendency to blame the laboratory is always present, which can lead to controversy and repeat analyses and other problems.

Therefore, even if sampling is not part of the laboratory's assigned tasks, laboratory personnel should be at least aware of the sampling methods used and should have input into the sampling process. In fact, the laboratory manager or supervisor should refuse to run samples that have not been taken properly, since the analysis of such samples is essentially wasted effort. At the very least, the chemist's reservations about such samples must be noted on the analytical report.

SAMPLING PLANS

All sampling begins with a well thought-out, written sampling plan. Except in the simplest situations, such as sampling a well-stirred, homogeneous liquid, the sampling plan is a collaborative effort, involving laboratory personnel, sampling personnel, and others who may have special knowledge of problems likely to be encountered, or technical aspects of the material being sampled that could bear on the taking of samples. Sampling occurs outside the laboratory and may well involve chemical and physical considerations, as well as practical problems that are beyond the ken of the analytical chemist. The plan is detailed, especially with regard to when and where the samples are taken, who takes the samples, and the procedure for obtaining the samples.

A written procedure is necessary for a number of reasons.

1. Writing a sampling plan forces constructive thinking on the part of the writer and his or her collaborators. Putting words on paper forces thought to embrace logic. Also the written plan may stimulate others to offer suggestions and criticism, which may not surface in an oral discussion of sampling.

2. A written plan avoids misconceptions and misunderstandings if it is written clearly and unambiguously. Oral instructions can be

misunderstood, or translated in the hearer's mind into something other than intended. Also, if personnel changes occur, the written plan stands as instructions as to how the sampling is to be accomplished.

3. The written plan offers guidelines and standard operating procedures (SOPs) to sampling personnel to ensure that sampling is accomplished as planned and desired.

4. The written plan provides documentation for quality assurance purposes that the samples were taken correctly and thus that there should be no quality problems based on sampling.

The remainder of this chapter will be concerned with the various factors that should be considered in writing a sampling plan.

Objective

The first thing to consider in writing a sampling plan is the objective or purpose for taking the sample and running the analysis. It is important that the client be contacted and his or her needs be established as unambiguously as possible, since this will define the objective of the program.

The "classical" situation is where the analyst is confronted with a large batch of solid material and needs to analyze a small quantity to determine the chemical composition of the batch. A "representative" sample is needed, that is, one that is as close as possible to the chemical characteristics of the batch.

However, there are many situations that do not correspond to this classical case. For example, the objective may be simply to determine if a contaminant is present at any concentration in the batch. It may then be necessary to sample those areas of the batch most likely to contain the contaminant, such as the material contacting the side walls or bottom of the container. The objective may be to measure the variation of the analyte with time as in measuring the pollutants in the waste stream of a manufacturing plant, where the bulk flow of the waste stream as well as the contaminant level may vary with time. Another example is the determination of exposure of personnel in a manufacturing plant to a toxic chemical, where it is necessary to measure the concentration of the chemical in the breathing zone of the employee over an 8-hour time period rather than the concentration in the atmosphere.

Defining the objective of the analysis in as clear and unambiguous a manner as possible is the first step in devising a sampling plan because all other factors in the plan depend on it.

Nature of the Matrix

Another important factor in the sampling scheme is the type and characteristics of the matrix to be sampled. Solids, liquids, and gases will each require different sampling techniques. Another factor may be the boundaries of the material. Is the matrix fixed in time and space, as in a simple lot of bulk material, or are the boundaries vague, as in the air surrounding a manufacturing plant, or is the matrix varying in both volume and composition, as in the wastewater plant discharge?

The homogeneity of the matrix is a very important factor in the sampling plan. Solids tend to be less homogeneous than liquids or gases, but even in the latter, layering can be present. Not just the fact of inhomogeneity, but its nature should be understood. It is known that a solid pile will tend to have smaller particles on the bottom of the pile than on the top because the small particles tend to fall through the holes between the larger particles, and these small particles may have different chemical composition than the large ones. In liquids and gases denser material will tend to concentrate in the bottom layers of the material.

The sampling plan indicates whether steps must be taken to eliminate inhomogeneity, such as mixing, or whether the measurement of inhomogeneity is necessary or even a major objective of the analyses.

The matrix may be multiphase, such as solid/liquid, liquid/gas, solid/gas, or all phases existing simultaneously. It may be necessary to decide whether the analyte should be measured in the total sample or in one or more of the separate phases. In the former case, mixing techniques may need to be described, while in the latter separation techniques may be necessary. Measurement of the volumes of the different phases or relative volumes may be necessary.

Type of Sample

In general terms there are three types of samples to be considered in a sampling plan. The first of these is the representative sample, or one that attempts to simulate the composition and properties of the bulk material as closely as possible. Such a sample may be a composite of many subsamples or increments, which are then thoroughly mixed prior to analysis.

The second type of sample is the selective sample, where the objective is to sample from those areas of the bulk material where the analyte of interest is most likely to be found. Examples would be analysis of food for rodent hairs or excrement, analysis of air in the breathing zone of a worker, or analysis of a metal for corrosion products.

The third type of sample could be called a protocol sample, which is a sample taken according to a previously agreed upon sampling procedure. Government regulations, contracts between supplier and purchaser, and other similar agreements often specify the kind of sample to be taken and the procedure to be used.

Statistical Considerations

Statistical considerations come into play in designing a sampling plan when it is known or suspected that the bulk material or population being sampled is not homogeneous. Depending on the objective of the analysis, it may be necessary to determine the degree of inhomogeneity of the bulk material, or it may simply be necessary to determine the overall average concentration of the analyte.

If the nature of the inhomogeneity is known, such as in the case of layering in a liquid or solid, the sampling plan may consist of taking samples at various levels in the population. This will provide information on the variation of the analyte concentration with one or two dimensions in the bulk of the material.

On the other hand, if the inhomogeneity itself is random, as may happen in soil or agricultural products, it is necessary to obtain a random sample of the population and apply statistical reasoning. To obtain a random sample, the bulk material is either physically or conceptually divided into discrete areas, zones, or units that are consecutively numbered. Samples are then taken according to a series of random numbers, either from a table of random numbers or from a random number generator such as a computer program. The objective of the program is to ensure that each zone or unit sampled has the same probability of being sampled as every other zone or unit.

The resulting samples, or more properly, increments, may be analyzed separately or composited for an overall average analysis of the analyte. In the former case, the variation due to inhomogeneity may be calculated, using Equation (2-7), provided the variation due to the measurement process is known, and, of course, the overall concentration of analyte can be calculated by averaging the individual analyses.

Although from a theoretical point of view the random sample approach is the best for statistical analysis of results, in practice it is often difficult to apply, unless the bulk material is relatively small in size. Random sampling from the interior of large quantities of bulk material, such as boxcars, can be impossible from a practical point of view. On the other hand, if the population is a moving stream, such as a fluid or a manufacturing production line, time may be used instead of a geographic zone, and samples taken at random times.

One question that occasionally arises is how many samples should be taken and averaged to obtain a predetermined level of uncertainty in the final average. Provided the measurement uncertainty is known, from previous analyses, Equation (2-5) enables the calculation of the standard deviation of a mean value, which can then be used to calculate 95 or 99 percent confidence limits. Note, however, that the standard deviation of the mean is inversely proportional to the square root of the number of samples, which can lead to problems of cost if highly precise measurements are desired.

Another important application of statistics in sampling concerns the precision or confidence range the client needs to solve the problem or answer the question for which the analyses are being run. This may force the choice of analytical method, or the number of samples to be run and averaged to achieve the desired confidence range. Given the standard deviation of the single determination and the desired standard deviation of the mean, Equation (2-5) may be solved for N, the number of analyses needed per sample:

$$N = \left(\frac{S}{S_{\bar{x}}}\right)^2$$

Example: An analytical method has a standard deviation of ± 0.25 percent and a client needs to have a precision of ± 0.1 percent. To achieve this degree of precision the following number of analyses must be run per sample:

$$N = \left(\frac{0.25}{0.1}\right)^2 = 6.25 \quad \text{or} \quad 7 \text{ analyses}$$

The sampling plan must allow for sufficient sample to run the necessary analyses.

Sampling Techniques

The techniques of sampling, or methods and equipment used in sampling, can vary considerably, from simple "grab" sampling of homogeneous solids or liquids to the use of sophisticated automatic sampling equipment as found in some environmental analyses or in industrial hygiene work.

Sampling techniques are often matrix-specific and traditional, especially in the case of common materials such as agricultural products, ores, coal, and the like. In these cases it is prudent to use the traditional techniques, because interpretation of the analytical results by the laboratory's clients is easier if the sample is taken in the traditional way. Frequently a large body of knowledge exists to assist in interpreting results, based on a specific sampling technique. Using a new or unusual technique, even if demonstrably better in some way, may render this body of knowledge of doubtful value.

An example of a sampling technology that may be as sophisticated as an instrumental analysis is the sampling of wastewater streams. If the stream flow is constant, the automatic sampler may be set to take samples at regular periodic intervals over 24 hours to get an overall analysis of the waste discharge. On the other hand, if the flow is not constant, a flow-proportional sampler may be necessary to take more samples when flow is high and fewer when flow is low.

Another similar example is the analysis of air in the breathing zone of a worker. In this type of sampling, a constant-flow pump is used to draw air through a collecting device such as a tube of absorbent material or a filter membrane, during an 8-hour working day.

In both of these types of analysis, and in other similar cases, sampling requires a high degree of skill and technical knowledge on the part of the sampler. Equipment must be accurately calibrated before and after analysis and monitored frequently during the sampling period to ensure correct operation. For good quality assurance and especially for legal defensibility in forensic situations, this kind of sampling is best done by persons with documented training in the techniques and preferably by degreed analytical chemists.

All personnel involved in sampling must receive some training in the technique to be used, no matter how simple it is. Document such training and write down the sampling procedure. A copy of this written procedure should be available to the sampler, preferably at or near the place where the sample is taken.

Apply labels to the samples as soon as they are taken. The label should show the date, time, and place of taking the sample, and the name of the sampler, preferably his or her signature. Give a sample number to the sample at this time, even if this is only a dummy number used to distinguish one sample from another. Take care that the labels are not easily destroyed or exchanged between samples. (For example, never place the label on the cap of the container, since caps can be exchanged inadvertently.)

Sample Preservation

In many cases the analyte is unstable, decreasing with time of exposure to air, light, the walls of the container, or high temperature. This will require that the sample be preserved by refrigeration, freezing, blanketing with inert gas, addition of acid, or the use of opaque or brown glass containers.

Determine the need for preservation before the sampling plan is written. This may require a research project to determine the rate of analyte deterioration under various preservation techniques. Part of this project should be the determination of the maximum holding time for the samples for accurate analytical results. Taylor[1] recommends that the maximum holding time not exceed the time for the analyte concentration to decrease by an amount equal to three times the standard deviation of the analysis, for a 95 percent confidence in the result.

One important consideration in sample preservation is the need to avoid contamination in the sample. This is especially important in analyses involving trace quantities of analytes, such as in most environmental work. Here it may be necessary to take field blank samples and trip blank samples to ensure that samples have not been contaminated by simple exposure in the field or in transit, or that contamination can be evaluated.

It is possible to purchase sample containers that are guaranteed not to be contaminated, and this may be a viable option in some cases of trace analyses. It may also be worthwhile to develop special cleaning techniques for sample containers, such as baking at 400°C to destroy organic contaminants. The type of analyte may also dictate the type of container to use, such as glass or plastic.

Special sampling techniques may also be needed to preserve samples. An example is the use of Teflon™-lined caps and filling the container to the very top to avoid headspace when sampling water for volatile organic analytes.

Also, give some consideration to preserving samples in transit between the sampling spot and the laboratory if considerable time or distance intervenes. Samples may need to be packed in ice or dry ice in heat-insulated boxes, and special packing materials may be needed to avoid breakage. A broken sample bottle not only voids that analysis, but may contaminate other samples as well.

Laboratory Sample Handling

The sampling plan should also define necessary handling procedures once the sample is received in the laboratory. Log in samples to the laboratory system as soon as possible after being received, and in no case log in more than 24 hours after receipt. In the case of forensic samples or for strict legal defensibility, begin a chain of custody procedures. This will require documentation that the sample has been received, by whom the sample was delivered, and the person who received and logged the sample. It is important that a "hard copy" record be made of sample receipt, that is, a computer memory or disk storage is not sufficient.

After log-in, samples are usually stored before analysis. Follow preservation precautions during storage as well as in transportation. The basic principle is that the sample, when analyzed, should have the same chemical and physical characteristics as when sampled.

Write down sample pretreatment steps prior to analysis, such as grinding, compositing, mixing, subsampling, either as part of the sampling plan or part of the analytical method.

Finally, in many cases it is necessary to take special steps in disposing of samples after analysis. This may be necessary if the samples are especially hazardous or toxic, or in the case of forensic samples, to ensure that the sample is not disposed of prematurely, since further analysis may be necessary. Forensic samples may require "bonded storage" if they may need to be produced for legal purposes.

DOCUMENTATION FOR QUALITY ASSURANCE

As part of the laboratory operations leading to an analytical result, the sampling steps should be thoroughly documented for quality assurance purposes. The following are documents that should be available and filed:

1. The sampling plan itself. This should be as detailed as possible, and available to both sampling and laboratory personnel.
2. The sampling procedure(s). This may be part of the sampling plan or may be in the form of standard operating procedures referenced in the sampling plan.
3. Sample-handling procedures. This may also be part of the sampling plan or separate SOPs.
4. Documentation of the training of sampling personnel. These training records may be kept as part of the employee's personnel records, or as part of the project file.
5. Sampling audit reports. Audit the sampling process periodically, by quality assurance personnel or laboratory supervision. The auditor should follow the sampling procedure, checking for compliance with the written sampling plan. Note any deviations from the written plan. Write a report of the audit and submit it to the

quality assurance officer who will take steps to resolve any deviations and retain the report on file.

REFERENCE

1. Taylor, J. K. *Quality Assurance of Chemical Measurements*, Chelsea, Mich.: Lewis Publishers, Inc., 1988.

CHAPTER

5

Analytical Methods

The comment has frequently been made that one of the major sources of confusion in the world is poor communication. Communication requires at least three components: the person transmitting the information, the person receiving the information, and the medium of transmission. In the analytical laboratory the analytical method is the medium by which management informs the workers of the procedures to be used in performing the work. Looked at in this way, the analytical method acquires an importance and respect not often accorded to it in the modern analytical laboratory.

In many laboratories the methods used are not written down, or if they are, they are in a very sketchy form. New workers are simply trained in the methods used by experienced workers, with little reference to a written copy of the method. In many laboratories procedures are simply outlined on index cards, written on scraps of paper, or penciled in margins of textbooks or notebooks.

The hazards of this type of operation should be obvious. If methods are not written out and available to the analysts, the possibility of intentional or unintentional variations in the method is always present. Technicians (nondegreed personnel), since they do not usually have the scientific background to completely understand the reasons for various operations, may be tempted to take shortcuts, eliminate steps, substitute reagents, and so on, sometimes in a commendable spirit of increasing productivity. Such "creative technicianship" may lead to loss of accuracy, precision, or both.

In addition, unless the method is written clearly and unambiguously, even professional chemists can misunderstand instructions, leading to poor results. This is one of the major reasons for wide variation in "round-robin" evaluations of analytical methods involving multiple laboratories, and one of the main reasons for conducting such evaluations: to be sure that the written method is thoroughly understandable by chemists.

Since management is ultimately responsible for the quality of analytical results it is important that workers understand the importance of the written method. There are three general principles involved:

1. All methods used in the laboratory should be in written form. The methods should be collected in a "methods manual," and available to all workers at all times.
2. All methods used in the laboratory should be authorized for such use by management, and files kept of documents indicating that each method has been so authorized.
3. All employees should be informed that unauthorized methods may not be used in the laboratory.

SOURCES OF ANALYTICAL METHODS

The proper choice of an analytical method is subject to many considerations. The method must first of all be consistent with the level of analyte being measured and must be appropriate for the matrix that will contain the analyte. Other factors to be considered are availability of instruments and equipment, speed with which results are required, cost, convenience, safety factors, and, of course, accuracy and precision. Most of these factors represent clients' needs, while others are dictated by practical considerations of the laboratory's operations.

There are, in general, four sources of analytical methods. These are standard methods, official methods, literature methods, and in-house developed methods.

Standard Methods. By far the best methods available in the analytical literature are those methods developed by standard-setting organizations such as the American Society for Testing and Materials (ASTM) or the Association of Official Analytical Chemists (AOAC). One of the major reasons for existence of these organizations is the development of test methods of proven accuracy and precision. These methods are subjected to intensive investigation by many individuals and laboratories before being awarded the status of "standard," and are therefore usually the best that can be had.

There are hundreds of standard-setting organizations in this country and worldwide, many of them producing books of methods that are internationally recognized as the methods of choice for a particular analysis or a specific matrix. Wherever possible standard methods should be used in the laboratory because they usually have been thoroughly tested, are widely accepted, and have associated with them a body of knowledge regarding accuracy, precision, interferences, and so on. Because of the long time required to establish a standard they may occasionally appear to be somewhat outdated, but it is always better to be safe than sorry in choosing a method. Standard methods usually require little or no validation work by the laboratory, unless modifications are made or they are applied to matrices other than those for which they were developed.

Official Methods. The term "official method" may be applied to methods mandated to be used by government organizations such as the Environmental Protection Agency (EPA) or the National Institute of Occupational Safety and Health (NIOSH) in cases of analysis for compliance with government regulations. Because of their importance and careful scrutiny by corporations and other organizations being regulated, these methods are generally thoroughly validated before being released for use. Since laboratories involved in this kind of work have no choice, there is little need to validate these methods. (Government agencies usually will not accept modifications of these methods by laboratories without a great deal of validation work by the laboratory.)

Literature Methods. The literature on analytical chemistry is extensive. Not only are there general-purpose analytical journals such as *Analytical Chemistry* and the *Journal of the AOAC,* but there are numerous specialized journals such as those dealing with food chemistry, pharmaceuticals, plastics, metals, paper, and so on, which publish analytical methods. The patent literature and house organs, especially of instrument manufacturers, are also sources of analytical methods.

Any analytical method abstracted from the literature should be used with caution. The authors of the original article are often far from unbiased in their assessment of the usefulness, convenience, accuracy, or precision of their "brainchild." Therefore a thorough validation study should be made of any such method.

In-House Developed Methods. The same precaution applies to methods developed in-house. No matter how simple or straightforward, run and document a thorough validation study before the method is authorized by management and used on client samples in the laboratory.

EVALUATION OF ANALYTICAL METHODS

Validation Studies

The term "validation" has been used in the analytical literature with several different meanings. For the purposes of this book the term is used to mean the process of establishing the suitability of a method for authorization for use in the laboratory. A validation study is a series of experiments run to evaluate the method for such suitability.

Standard methods and official methods, as previously defined, generally need only a minimum of work to demonstrate their suitability for authorization. Standard methods do not become standards until a great deal of validation work has been done, and official methods are required to be used whether the laboratory does a validation study or not. Nevertheless, run a few analyses using reference samples or spiked samples to be sure the method is well written and acceptable results can be obtained.

Subject methods adopted from the literature, or in-house developed methods, or major modifications of existing methods to a validation study before authorizing them for routine laboratory use.

Write out the method explicitly and then examine the written procedure for any ambiguity that may be present. The validation study should be carried out by an experienced analyst to avoid any chances of errors due to analyst inexperience or learning curve. The purpose of the study is to evaluate the errors and possible bias in the method itself.

Obtain, or prepare by spiking, a series of four or five samples covering the range of concentrations likely to be encountered in unknown samples. The matrix should be the same as that which will be in the unknowns, and, in the case of spiked samples, should exhibit zero concentration of the analyte before spiking. Take proper precautions to ensure that the spike is homogeneous, and that it is in the same physical and chemical state as the analyte would be in an unknown sample.

Run replicate analyses (four or five) on each of the samples, using the method exactly as written. Run the replicates in random order, and on different days, to subject the method to whatever uncontrolled or unrecognized variables may be operating. The same analyst should run all replicates to eliminate between-analyst variations.

When all analyses have been run the data may be evaluated. First calculate the mean and variance for each sample, using the replicates for that sample. The variances for the samples representing the highest and lowest concentration levels may be compared, using the F test (Chapter 2) to determine if there is a significant difference. If the F test indicates that there is a difference, a plot of standard deviation versus concentration may be constructed to determine if the standard deviation is constant or proportional to the concentration level. In the latter case, the coefficient of variation or relative standard deviation is constant, and averaging the coefficient of variation or relative standard deviation over all samples will give an estimate of the precision of the method.

If the F test shows no significant difference between the standard deviations of the highest and lowest concentrations, it can be concluded that the standard deviation is constant. In this case the variances may be pooled, and an overall standard deviation calculated from

$$S^2 = \frac{\sum N_i S_i^2}{\sum N_i - k} \tag{5-1}$$

where S is the standard deviation, N_i is the number of replicates of the ith sample, S_i^2 is the variance of the ith sample, and k is the number of samples analyzed, that is, the number of concentrations used.

Next construct a graph by plotting the means of the replicate analyses on the Y axis against the known concentrations of the samples on the X axis. It is important that this choice of axes be used, since the regression line (or "least squares" straight line) will be calculated for the plot, and reversing the axes will give an incorrect equation for the line.

To calculate the least squares line proceed as follows. Assume there are N pairs of data, x, y, where the x's are the known concentrations of the samples we ran, and the y's are the "found" concentrations, that is, the means of the replicates for the samples. Calculate the following:

$$S = \sum x \qquad Q = \sum x^2$$

$$Y = \sum y \qquad L = \sum y^2$$

$$P = \sum xy \qquad D = NQ - S^2$$

Then the slope of the least squares straight line will be given by

$$m = \frac{NP - SY}{D} \tag{5-2}$$

and the y intercept (i.e., the value of y at $x = 0$) by

$$b = \frac{QY - SP}{D} \tag{5-3}$$

These two results, *m* and *b*, are the parameters of the equation of a straight line:

$$y = mx + b \tag{5-4}$$

which is the least squares straight line. (The name comes from the fact that this line represents the minimum sum of the squares of the distance from the line to the plotted points, measured in the *y* direction.)

Theoretically, if there were no bias or error in our measurements, the least squares line would be a line with a slope of 1, and an intercept on the *y* axis of 0. However, this seldom is the case, and the question then arises, given an intercept *b* different from 0 and a slope *m* different from 1, what is the probability that these represent merely a random variation from a true slope of unity and intercept of zero? If the experiment were repeated there is a high probability that the new slope and intercept would be different from the first one.

Mandel and Linnig[1] addressed this question in a paper published 30 years ago. They showed how to construct a confidence band around the least squares straight line, corresponding to any degree of confidence, for example, 95 percent or 99 percent. If the theoretical line of zero intercept and unity slope lies within the confidence band of the experimental line, then one can assume with the given level of confidence that the experimental slope and intercept are not significantly different from unity and zero, respectively.

The confidence band of Mandel and Linnig consists of two arms of a hyperbola, which may be calculated as follows. First calculate the following expressions:

$$s^2 = \frac{1}{N-2}\left[L - \frac{y^2}{N} - \frac{D\,m^2}{N}\right] \tag{5-5}$$

$$K = (2Fs^2)^{1/2} \tag{5-6}$$

where *F* is the *F* factor determined from an *F* table, giving *F* at the desired confidence interval and entered with 2 (for the numerator) and $N - 2$ (for the denominator) degrees of freedom. The two arms of the hyperbola are then given by the equation

$$y = b + mx \pm K\left\{\left(\frac{1}{N}\right)\left[1 + \frac{N^2(x - \bar{x})^2}{D}\right]\right\}^{1/2} \qquad \textbf{(5-7)}$$

where $\bar{x}$ is the average of the x values.

Although Equation (5-7) seems somewhat formidable, it is relatively easy to handle with a personal computer or even a hand calculator. From Equation (5-7) two values of y can be calculated for a series of assumed x values, one corresponding to the + sign and the other to the − sign. When plotted on a graph the positive sign values will correspond to the upper arm of the hyperbola and the negative sign values to the lower arm. The following example and Figure 5-1 (page 68) show the calculations and resulting plots.

Example: An analyst performing a validation study prepares five samples of known concentration and analyses each sample four times. The x values in the table below represent the known concentrations of the samples, and the y values are the means of the four analyses of each sample.

	x	y	xy	x^2	y^2
	1.00	0.95	0.95	1.00	0.903
	2.00	2.30	4.60	4.00	5.290
	3.00	3.00	9.00	9.00	9.000
	4.00	3.75	15.00	16.00	14.063
	5.00	5.05	25.25	25.00	25.503
Sum	15.00	15.05	54.80	55.00	54.75

Therefore,

$$S = 15.00 \qquad Q = 55.00$$

$$Y = 15.05 \qquad L = 54.75$$

$$P = 54.80 \qquad D = 5(55.00) - (15)^2 = 50.00$$

The slope m and intecept b are then given by

$$m = \frac{5(54.80) - 15(15.05)}{50.00} = 0.965$$

$$b = \frac{55.00(15.05) - 15.00(54.80)}{50.00} = 0.115$$

and the equation of the least squares line is

$$y = 0.965x + 0.115$$

To calculate the 95 percent confidence hyperbola, first calculate s^2 and K as follows:

$$s^2 = \frac{1}{3}\left[54.75 - \frac{(15.05)^2}{5} - \frac{50.00}{5}(.965)^2\right]$$

$$= 0.047$$

F for 95 percent confidence with 2 and $5 - 2 = 3$ degrees of freedom is 16.0, therefore,

$$K = [2(16.0)(0.047)]^{1/2} = 1.22$$

and the equations for the hyperbola are

$$y = 0.115 + 0.965x \pm 1.22\left\{0.2\left[1 + (x - 3.0)^2\,\frac{25}{50}\right]\right\}^{1/2}$$

where 3.0 is the average value of x (i.e., $\frac{15}{5}$).

The graph can now be examined to determine if the theoretical line of unity slope and zero intercept falls within the confidence band established by the two branches of the hyperbola. If the line is well centered there can be little doubt that the method is unbiased. If the line is not well centered but falls close to the limits, it might be wise to recalculate, using a tighter confidence limit (say 99 percent rather than 95 percent).

Comparison of Analytical Methods

This technique for validating new methods can be extended to the comparison of a new method with an old, established method. The

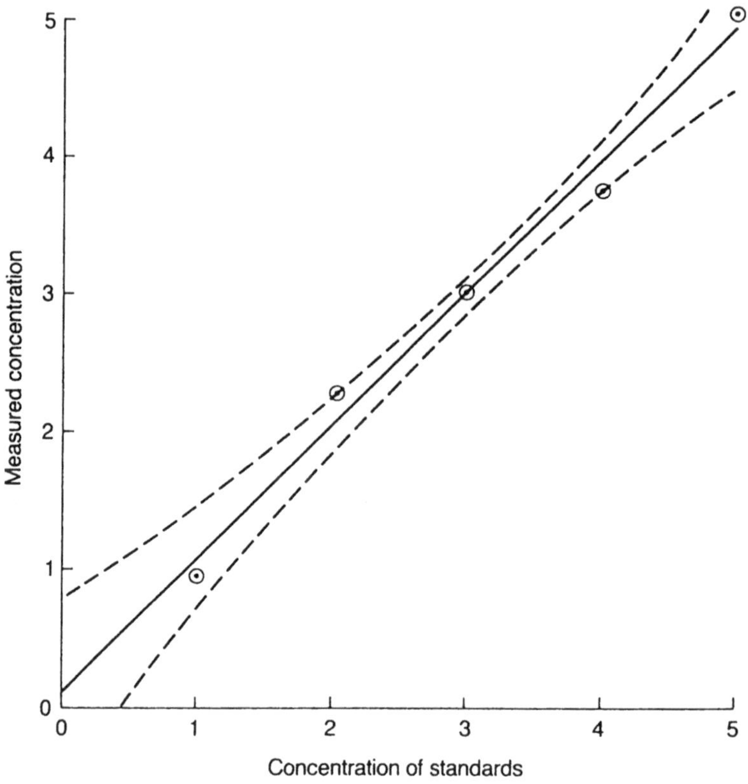

Figure 5-1. Plot of data for example.

new method may be totally new or a modification of the old method. In this case it is not necessary to prepare samples of known concentrations. It is only necessary to obtain a series of samples covering the concentration range of interest. The analyses by the old method are assumed to represent the true concentrations of the samples and are used for the x coordinates on the plot. Analyses by the new or modified method are assigned to the y coordinate, and the calculations performed as in the previous example. If samples are analyzed in replicate by both methods, the standard deviations can be compared by the F test to determine if there is a significant difference in imprecision. As in the case of validation studies, the least squares straight line on a plot of the new results versus the old should theoretically have a slope of unity and an intercept of zero, but it may be necessary to plot the confidence band to determine if this is the case.

Ruggedness Testing of Methods

Analytical methods of any degree of complexity involve a series of operations carried out on a sample and a final measurement step. These operations often involve setting a series of "variables." For example, sample size, temperature and time of drying, filtering, filter medium used, time of shaking an extraction may be considered variables in the analysis. Some methods may be sensitive to minor variations in these variables, while others may not. The latter type have been called "rugged" by Youden in Youden and Steiner[2] and he has described a procedure for evaluating the ruggedness of a method. It is recommended that all new analytical methods be subjected to a Youden ruggedness test before being accepted as an authorized method for a laboratory.

The Youden ruggedness test is an efficient one, since the effect of seven variables can be evaluated with only eight analyses of one sample. The first step is to determine the seven variables to be evaluated. Examine the written method and identify those steps that could possibly affect the final result. In addition to those given above others might be solution volume, reagent concentration, pH, rate of stirring, time and temperature of heating, and so on. If seven variables cannot be identified, one or two dummy variables may be used.

Assign seven capital letters *A* through *G* to one level of each of the variables, and the corresponding lowercase letters *a* through *g* to the other level of the variables. For example, if the method calls for 1-gram samples the letter *A* might correspond to 1.2 grams and *a* to 0.8 gram. *B* could be one type of filter medium and *b* another, and so on. The analyses can then be carried out according to the scheme given in Table 5-1. In the table each column corresponds to a single analysis with the variables set according to the letters as shown. The results correspond to the letters *s* through *z* given in the final row.

From the results the effect of each of the variables can be cal-

Table
5-1. The Youden Ruggedness Test

	Combination or Analysis Number							
Factor Value	**1**	**2**	**3**	**4**	**5**	**6**	**7**	**8**
A or *a*	*A*	*A*	*A*	*A*	*a*	*a*	*a*	*a*
B or *b*	*B*	*B*	*b*	*b*	*B*	*B*	*b*	*b*
C or *c*	*C*	*c*	*C*	*c*	*C*	*c*	*C*	*c*
D or *d*	*D*	*D*	*d*	*d*	*d*	*d*	*D*	*D*
E or *e*	*E*	*e*	*E*	*e*	*e*	*E*	*e*	*E*
F or *f*	*F*	*f*	*f*	*F*	*F*	*f*	*f*	*F*
G or *g*	*G*	*g*	*g*	*G*	*g*	*G*	*G*	*g*
Results	*s*	*t*	*u*	*v*	*w*	*x*	*y*	*z*

Source: Association of Official Analytical Chemists. Reprinted by permission of the publisher.

culated by averaging those four analyses that contain the variable in the capital level and those containing it in the lowercase level. For example, the results *s*, *t*, *u*, and *v* all contain the *A* level of the first variable and *w*, *x*, *y*, and *z* all contain the *a* level. The other six variables are all present in these results twice at the capital level and twice at the lowercase level, and hence their effects will cancel out. Subtracting $(w + x + y + z)/4$ from $(s + t + u + v)/4$ will show the effect of variable *A*. Similarly, the effect of the seventh variable, that is, changing from *G* to *g* is given by the difference between $(s + v + x + y)/4$ and $(t + u + w + z)/4$. Again the effects of the other variables cancel, as will be true in the case of each variable if the correct choice of results is made.

Collect and examine the seven average differences *A*–*a* through *G*–*g*. Any variable with a real effect will have a difference considerably larger than the others. Give special attention to these variables in the final written version of the method, spelling out the extent to which they must be controlled to produce quality results.

One by-product of the Youden ruggedness test is that the standard deviation of the results *s* through *z* is an excellent measure of the imprecision to be expected in routine analyses, since the test deliberately introduces the type of variation in variables that might be expected to occur in normal use.

WRITING ANALYTICAL METHODS

All methods used in the analytical laboratory should be in written form and kept in the laboratory in an area where they are accessible to all analysts. The methods may be kept in a methods manual of the loose-leaf binder type. Photocopies may be made for the use of individual analysts if necessary.

The methods manual should contain only those methods currently in use. Remove obsolete methods and relegate them to a "historical file" of methods so that they are available for tracing the method used at a given time and date of analysis. Keep the historical file in the Q.A. office and not in the laboratory, to prevent analysts from using the wrong method.

The method as written should correspond exactly to the way the analysis is run. Laboratory supervisors should monitor operations carefully to be sure that deviations from the written method are not occurring. If not monitored, slight variations in the methods can creep in as analysts attempt to "improve" the methods. Often this is done from laudable motives, such as to increase the speed or productivity of the method or for safety reasons. Such deviations are not necessarily bad and may, in fact, represent improvements in the method. However, rigorously evaluate such modifications of methods to be sure that they are not deleterious to quality, and then incorporate them into the written method. Clearly distinguish the new or modified method from the old by a new method number, and remove the old method from the laboratory method manual and place it in the historical file.

Format for Analytical Methods

Many different formats are used for written analytical methods. Regardless of the actual format used, it is prudent to have a format outline to which the writers of methods can refer. This will provide a logical uniformity to the methods in use and ensure that all necessary information is included. It will be especially helpful to the writers charged with reducing the method to written form.

The following is a format that the writer can recommend based on practical experience with it. Each part of the method is listed in the order in which it should appear. Do not leave out any section. If the information for that section is not available, explicitly state that fact.

Format Outline

Number of the Method. Assign a number to all methods authorized for laboratory use for traceability purposes. Note the number in the analyst's notebook and on the analytical report. Methods may simply be numbered consecutively or according to any logical system, for example, the use of an initial letter to designate the substrate (*F* for food, *W* for water, and so on).

The important point is that there should be only one number for any given method and only one method for any given number. Although this seems obvious, experience shows it is frequently not followed. For example, there may be a temptation to give a blanket number to a series of methods that may differ only slightly from one another. This is likely to cause confusion when tracing a result to the method used. The reverse of this, namely, giving different numbers to the same method may occur if two different departments in the laboratory use the same method. Coordination by the Q.A. officer can help prevent this.

Date Authorized

Title. Take care with the choice of title. It should be brief and contain the name of both analyte and matrix or substrate, for example, "Determination of *Nitrogen* in *Animal Feeds.*" If more than one method exists for a given analyte/matrix combination, the name may include a designation of the measurement method, for example, "*Spectrophotometric* Determination of *Color* in *Wastewater.*" Occasionally the method used will depend on the level of analyte present in the matrix and should be noted in the title, for example, "Determination of *Low-Level Phenol* in *Wastewater.*"

References. Include references to the literature on which the method is based, or in-house documentation of validation studies.

Scope. The scope shows the range of analyte concentrations for which the method is useful, the type and nature of the matrix to which it may be applied, and an estimate of the time for a single analysis. It also indicates interfering substances. In other words it should enable the analyst to decide quickly whether the method is suitable for the analytical problem with which he or she may be faced.

Basic Principles. This section describes the physical, chemical, or biological principles on which the method is based. The equations for unusual chemical reactions are given, separation processes briefly described, and the effect of interfering substances is noted. Procedural steps designed to eliminate or minimize interferences should be given.

Apparatus and Reagents. This section describes instruments and unusual apparatus that are required. Common laboratory equipment, for example, pH meters or analytical balances need not be described, unless special capabilities are needed, for example, the need to measure to 0.001 pH units, or micro or semi-micro weighing capability. Common glassware need not be listed, but specialized pieces, for example, a Soxhlet extractor should be mentioned.

Fully describe reagents, including chemical name, purity, and description of method of preparation for those which need to be prepared prior to analysis. Give shelf life where stability may be a problem.

Safety Precautions. Describe any safety precautions peculiar to the analysis. These might include necessity for working in a hood, steps to avoid hazardous reactions such as explosions, need for special safety devices or clothing, and special precautions to deal with hazardous waste or sample disposal. Give the reasons for the safety precautions so that the analyst may assess the degree of hazard represented.

Procedure. This is the heart of the method and should be written with special care. The author should put herself in the position of the reader, who may be a recent B.S. chemist given the method and asked to analyze a sample. The following guidelines may help in writing this section:

1. Follow a strict time sequence, exactly as the test is run. There is nothing more frustrating than to suddenly come upon an instruction such as "Add 20 mL of a 1:10 dilution of a 50:50 mixture of reagents A and B, which have previously been mixed and filtered." This means the analyst must turn his or her attention from the sequence of operations in the procedure to mix two reagents, filter the mixture, and dilute 1:10 before continuing.

2. Avoid abbreviations or jargon unless you are sure they are commonly understood. Remember that an abbreviation may be well understood by you and your staff, but not by a new employee. If convenient, an abbreviation may be defined the first time it is used and then used after that.

3. Be specific. Don't say "neutralize with HCl" when what you mean is "add HCl dropwise, to a pH of 7.00 ± 0.02." On the other hand, it is not necessary to spell out in detail common laboratory operations such as weighing or titrating.

4. Indicate critical steps in the analysis and the consequences if care is not taken.

5. Use short sentences and avoid convoluted phrases that can lead to misinterpretation. The author recently ran across the phrase "this method will detect not less than 0.02 ppm of . . ." in a method. The use of "not less than" was momentarily confusing until it was realized that it was equivalent to "more than," and the statement was meant to define a detection limit.

Calculations. Give the equation(s) necessary to calculate the result of the analysis, including the units of all variables and the units of derived results. If the equations are not straightforward, indicate how they were derived.

Statistics. Give all information available on precision and accuracy of the method in summarized form. Refer, if necessary, to the source of the method and any validation studies that were run in the laboratory.

Quality Assurance. Indicate what reference samples are available, and how often they are run. The need for calibration of instruments and equipment may be addressed and references given to calibration procedures.

Comments. Any special comments or remarks that may aid in the understanding of the method or interpretation of results should be inserted here.

AUTHORIZATION OF ANALYTICAL METHODS

As discussed at the beginning of this chapter, the set of analytical methods used in the laboratory is, in effect, the communication medium by which management controls the way the work is done. Management must thus exert some control over the methods that are used. In many laboratories this is accomplished in an informal way, if at all. Many managers have a simple faith in the competence of their laboratory supervisors and analysts, and in most cases this faith is well placed. However, in the modern world, where serious consequences may result from poor quality data, much more emphasis is being placed on the duty and responsibility of the manager for accurate and reproducible results, and the manager may find herself held personally responsible for poor data quality.

Therefore, it is incumbent on managers to become thoroughly familiar with the methods that are actually in use in their laboratories, with all the limitations and variability that may be expected. For quality assurance purposes it is necessary that documentation exists which demonstrates that the method has been authorized for use by the manager responsible for the quality of the data. This may be accomplished by establishing a formal procedure whereby the manager reviews all pertinent references and in-house documentation and completes an authorization form that permits the analysts to use the method. Attach a copy of the form to the written method, and keep the original in the Quality Assurance officer's files. An example of such a form follows.

METHOD AUTHORIZATION FORM

Title of method: Date:

Submitted by:

Analyte (substance being analyzed for):

Brief description of method:

Brief description of validation study:

a. Accuracy (standard error of residuals, or evidence of bias):

b. Precision (standard deviation or coefficient of variance):

c. Interferences:

d. Applicable concentration range:

e. Limit of detection (give basis):

f. Results of ruggedness test:

Validation study performed by:

Validation data in notebook no.:

Approved by ______________________________
Name Title

Date of approval:

The use of such a formal procedure, together with a management directive that states that only *authorized* methods may be used, will prevent the encroachment and use of untested, unproven methods in analytical operations. Point out to analysts that the use of unauthorized methods without express management permission will not only jeopardize their positions, but may make them legally responsible for the results of poor data.

REFERENCES

1. Mandel, J., and F. J. Linnig. Study of Accuracy in Chemical Analysis Using Linear Calibration Curves, *Analytical Chemistry*, 29:743–749 (1957).
2. Youden, W. J., and E. H. Steiner. *Statistical Manual of the Association of Official Analytical Chemists*, Washington, D.C.: Association of Official Analytical Chemists, 1975.

CHAPTER

6

Quality Assurance and Instruments, Equipment, and Materials

It was stated in Chapter 1 that quality control implies controlling all factors in the analytical system that can impact on the analytical result. It is obvious that three important factors in the system are the instruments used, the auxiliary equipment used, and the materials that go into the analytical operations. In this chapter we will consider those aspects of these factors that are involved in quality assurance, and especially quality control.

INSTRUMENTS AND EQUIPMENT

Under this topic instruments, which are devices that measure chemical or physical properties of material, auxiliary equipment, which are devices that are nonmeasuring but that affect analytical results, for example, ovens, muffle furnaces, and centrifuges, and glassware (including plasticware), are included.

There are three general principles that apply in considering the quality assurance aspects of instruments and equipment. The first is that the equipment should be capable of doing the job required of it. Although this seems obvious, there are situations where the laboratory, due to financial constraints, may be tempted to purchase equipment, including instruments, that are only marginally capable of performing as desired. Invariably, such equipment will lead to frustration and poor quality data.

The second principle is that all equipment should be kept in optimal condition for use as needed. This implies both preventive maintenance and control over the use of the equipment by laboratory personnel.

The third principle is that equipment should be frequently monitored and evaluated. This implies calibration.

Calibration of Equipment

There is some semantic confusion in the analytical literature about the terms "calibration" and "standardization." Originally the word calibration meant to check the response of an instrument to a material of known property, and perhaps apply a correction factor to the "calibration" marks of the instrument. Standardization, on the other hand, meant to adjust the response of the instrument to conform to the known properties of the material, so that the instrument would then read correctly. Thus a balance would be calibrated by weighing a known mass certified by an independent supplier of such masses and a correction applied to all future weights based on the measurement. On the other hand, a spectrophotometer would be standardized for a colorimetric test by measuring absorbance of a series of solutions of known concentrations and constructing a "calibration curve" or "standard curve" for measuring concentrations of unknown solutions. Since in everyday usage the two terms have become effectively synonymous, they are used interchangeably in this book.

Many complex instruments, for example, gas chromatographs, GC/MSs, and atomic absorption spectrometers, are in reality *systems* of components, rather than simple measuring devices such as pH meters or balances. Even a simple gas chromatograph may have an automatic sample injector, an injection port with associated temperature control, a carrier gas flow system, a chromatographic column with temperature controls (constant or programmed), a detector with its temperature control and associated electronic circuitry, and a strip chart recorder or electronic integrator. It is obviously not practical to calibrate such an instrument or rather all the components of the system and so we fall back on standardization of the system by measuring a series of solutions of known concentrations. Of course, even simple instruments such as UV/visible spectrophotometers are usually standardized before use as measuring devices, and may also be calibrated with a material of known absorbance at a known wavelength.

The one weakness in this procedure is that it is assumed we know the concentrations of the standard solutions. As any experienced analytical chemist knows, this is one of the major sources of blunders in the laboratory. A common procedure is to prepare a concentrated solution and dilute it to prepare a series of lower concentrations. In this case, if an error is made in the preparation of the first solution, all solutions are in error. Another common mistake is to use a standard material that is not of the necessary purity because of degradation, improper storage, or contamination.

To prevent these types of errors, prepare each standard solution from scratch, not diluted down from a concentrate, and run a second standard immediately after standardization to ensure no mistake was made. Of course, if control charting is practiced, any error in standardization should be apparent when the check sample is run.

Linear Calibration Curves

It is assumed in most analyses that the response of the instrument to the material of interest will be *linear*, that is, the relationship will be that of a straight line:

$$y = mx + b \qquad \textbf{(6-1)}$$

where y is the response of the instrument, x is the concentration of the analyte, and m and b are the slope and intercept of the straight line. Ideally, b should be 0, but because of error in measurement it seldom is.

In the usual standardization procedure a series of standard solutions is prepared, and the response of the system measured to each solution. In the not-too-distant past, most calibration curves were manually plotted, and the best visual straight line drawn on the plot.

In many modern instruments a built-in microprocessor calculates *m* and *b* according to the least squares equations given in Chapter 5 and displays the graph on a computer terminal. If the analyst is satisfied with the linearity of the plot, the instrument then uses Equation (6-1) to calculate concentrations of the unknown samples from the instrument response.

In some cases, a standard error of estimate is calculated. This is arrived at as follows. It is assumed that, associated with each response y_i, there is an error e_i such that

$$y_i = \hat{y}_i + e_i = mx_i + b + e_i \tag{6-2}$$

These errors e_i are assumed to be normally distributed with a standard deviation S_e, calculated from

$$S_e = \left(\frac{\Sigma e_i^2}{N - 1}\right)^{1/2} \tag{6-3}$$

where *N* is the number of samples. The individual e_i's may be calculated and then used to calculate S_e. If an unknown solution is now measured, giving an instrument response *y*, then its concentration will be given by

$$x = \frac{y - b}{m} \tag{6-4}$$

and its standard deviation by

$$S_x = \frac{S_e}{m} \tag{6-5}$$

While Equation (6-5) is not rigorous because of the possible error in *m* and *b*, it may be used in most cases for an approximation.

Unfortunately, manual calculation of *m* and *b* is usually not practical in routine measurements and so is seldom done. It would be desirable if computerized equipment would give a printout of these parameters, but this is also usually not done on present-day instruments. A control chart of *m* and *b* or S_e would be an excellent check of instrument condition. Failing this we must rely on X-charting of a reference sample to detect degradation of instrument response.

Calibration by Comparison with Another Method

In some instances it is possible to calibrate an instrument by comparing instrument response to concentrations of unknowns measured by an alternative method. The alternative method may involve a completely different type of instrument or may be noninstrumental. It is assumed that the alternative method is known to be accurate, or it may be a widely accepted method. An example is the calibration of dissolved-oxygen meters against the Winkler iodometric titration.

Frequency of Calibration

The frequency with which instruments and equipment should be calibrated is a function of several factors, such as the nature of the instrument and its ruggedness, the frequency of use of the instrument, its environment (vibration, dust, and so on), and the demands of the analytical methodology. Calibration schedules may be set up according to the frequency of use or according to elapsed time. If an instrument is not used very often or is of a particularly sensitive nature it may need to be calibrated every time it is used. Instrumentation used for trace analysis, which is often of a complex nature, such as gas chromatographs, atomic absorption spectrometers, and HPLCs may require standardization each time they are used. On the other hand an infrared spectrophotometer may only need to be calibrated once a month as long as nothing has occurred to cause misalignment. Calibrate pH meters, which are very sensitive, each time they are used, or once or twice a day if they are used often. In general, professional judgment is required when deciding calibration schedules, with the general principle that the more often, the better.

The importance of the sample may affect calibration frequency. If the data may become part of the evidence in a legal suit or other forensic situation, or if costly decisions may depend on its accuracy, all instruments and equipment used should be calibrated before and after analysis.

Calibration of Common Laboratory Instruments and Equipment

The following is a brief summary of calibration methods and guidelines for frequency of calibration for the more common laboratory instruments.

Analytical Balances. Check with a high-quality (at least class S) 50-mg or 100-mg weight. Calibration frequency is dependent on frequency of use, either daily or weekly. One word of caution regarding electronic balances, which are becoming ubiquitous in the analytical laboratory. These balances rely on integrated circuits ("chips") and associated circuitry for their readout display. In one instance in the author's experience the failure of a component led to inaccurate weights, but only after the instrument was operating for some time. The balance was calibrated every morning after a 20-minute warm-up, but only malfunctioned later in the day. It may be wise with these instruments to calibrate twice a day, once in the morning and once in the afternoon.

Volumetric Glassware. If the method requires highest accuracy use only Class A glassware. Class A glassware is guaranteed by the manufacturer to meet National Institute for Technology and Standards (NITS, formerly NBS) specifications for accuracy, and thus does not ordinarily need calibration. Other types of glassware may be used if lower accuracy can be tolerated, but should be stored separately from Class A to avoid unintentional use when Class A is needed.

Although generally Class A glassware does not need to be calibrated, twice in the author's career of more than 40 years he has found burets that had been miscalibrated in that one of the numbers on the scale was missing. In one case a 50-mL buret had a number missing in the 40- to 50-mL range, being labeled 43, 44, 46, 47, . . . , 50. Titrations that required 45 mL or more were incorrectly read, while those at lesser volumes were read correctly. A defect such as this can be very difficult to detect without a full-scale calibration, and the defective item can be used for a long time before the situation is discovered.

Glassware may be calibrated gravimetrically using water, or in the case of small volumes, mercury.

Ovens. Thermometers used in ovens should be NITS (formerly NBS) certified or calibrated against NITS-certified thermometers. Check oven temperature daily.

Furnaces. Calibration of muffle furnaces is difficult. Fortunately for most purposes accurate calibration is seldom necessary. Optical pyrometers, or high-temperature probes, may be used, but usually require that the furnace door be at least partly open, leading to low readings. High-melting inorganic salts may be used, choosing two whose melting points bracket the desired temperature. Semiannual calibration is usually sufficient.

Ultraviolet/Visible Spectrophotometers. These instruments are usually calibrated by the standard calibration curve method. However, absolute calibration is possible with various standards. NITS-certified glasses are available, and a widely used standard is a solution of 0.0040 g/L of K_2CrO_4 in 0.05*M* KOH, which has peaks in both the ultraviolet and visible range. A dilute solution of $KMnO_4$ may be used to check resolution via the two peaks at 526 and 546 nm.

Standardization by the standard-solution technique may need to be done each time samples are run, or less frequently, if many samples are run. Absolute calibration should be done semiannually or when the instrument is suspected of malfunctioning, or when maintenance work is done such as installing a new light source.

pH Meters. Calibration is carried out with standard buffer solutions, usually at pH 4.00 and 7.00 with pH 10.00 buffers used if alkaline solutions are frequently encountered. Calibrate pH meters each time used or daily or twice daily if used frequently.

If the meter is often used to measure the pH of solutions that are particularly "dirty," such as wastewaters, which may contain high concentrations of solid matter or oil and grease, it may be found that the glass electrode has become fouled and response is not linear. Two or even three buffers will not be sufficient to detect the nonlinear response at pHs other than those of the buffers. In cases where dirty solutions are frequently measured, perform a weekly calibration of the meter with a series of buffers covering pH 3 to 8. A response curve can then be constructed and used to correct readings. Table 6-1 indicates how a series of solutions can be prepared, covering pHs from 3 to 8 using measured volumes of just two reagent solutions.

Table
6-1. Preparation of Solutions of Varying pH

pH	Solution A, mL	Solution B, mL
3.0	15.89	4.11
4.0	12.29	7.71
5.0	9.70	10.30
6.0	7.37	12.63
7.0	3.53	16.47
8.0	0.55	19.45

Solution A: 0.100*M* citric acid.
Solution B: 0.200*M* disodium phosphate.

Infrared Spectrophotometers. A thin film of polystyrene is the most common standard used to calibrate these instruments. It is stable and contains a rich mixture of absorption peaks. Percent transmission and wavenumber of the peaks can be compared with previous runs to detect any deterioration in performance. Standardization should be done whenever the instrument has been subjected to unusual vibrations or jarring, for example, after it has been moved to a new location.

Atomic Absorption Spectrophotometers. These instruments, like many of today's complex instruments, fall into the class of those calibrated each time they are used with a set of standard solutions. Certified standard solutions (1 g/L), which have been checked against NITS standards, are available for various elements from chemical supply houses.

Conductivity Meters. These instruments are usually calibrated against standard KCl or NaCl solutions. Frequency of calibration will vary with frequency of use, but should be done at least weekly, if heavily used.

Gas Chromatographs and High-Performance Liquid Chromatographs. These are instruments that fall in the class of those which are calibrated via calibration curves each time a batch of samples is run. Prepare calibration standard solutions with materials of high purity (>99.9%), which are commercially available in most cases. Check calibration curves frequently against previous runs to detect any changes in instrument response. Use a secondary standard, independently prepared, to check the calibration curve.

Maintenance

Instrumentation and equipment is the most potent amplifier of productivity available to the manager of the analytical laboratory. Even though the cost of instrumentation has increased substantially over the years, it has not equaled the increase in labor costs, which is, and has always been, the major element in analytical costs. This has made the investment in more and more sophisticated equipment an attractive method of allocating financial resources. What is sometimes not appreciated by supervision and management, though, is the return on investment of a proper maintenance program for costly equipment.

In the modern laboratory equipment and instrument repairs are beyond the capabilities of most laboratory personnel. The analytical chemist no longer is expected to be knowledgeable in the electronic field, beyond the most basic concepts. This has given rise to the new profession of instrument technician, one who specializes in the repair and maintenance of instruments and other laboratory equipment. Large laboratories, or labs in large organizations, may have one or more instrument technicians on the payroll, but most laboratories are forced to rely on outside help, either from the manufacturer or a firm that specializes in this service. The cost of this service is high, with hourly rates of $50 to $80, plus travel costs. Service contracts are available from most manufacturers and are probably a good idea, at least in the first year of an instrument's life. Preventive maintenance can do much to lower the cost of instrument repair.

Proper preventive maintenance begins with the environment in which the instrument is placed and used. Laboratory instruments are sensitive devices designed to give maximum response for a minimum input of detected material. For this reason they are vulnerable to environmental influences such as dust, vibrations, corrosive fumes, radiation, excessive heat, and humidity. If at all possible keep and use instruments in a separate room or rooms that are set aside solely for this purpose. At a minimum such a room must be air-conditioned in the summer and heated in the winter. It is desirable for the instrument room to have a heating, ventilation, and air-conditioning

(HVAC) system separate from the rest of the laboratory. If possible keep doors closed and the instrument room at a slightly positive air pressure, with respect to the environment outside the room. This will guard against the accidental intrusion of dust, fumes, and vapors from the lab or outside environment. With the very low detection limits achievable with some of today's instruments, it is mandatory to isolate these instruments from outside contamination for reproducible results. Simply opening a bottle of volatile material in the room can cause contamination problems for days to come.

Instruct personnel in the proper use and care of instruments. Prohibit employees who are not authorized to use instrumentation, that is, who have not been specifically instructed and trained, from using or handling costly instruments. Most people will approach a new instrument with caution, being mindful of the cost of the equipment and the dire consequences of failure, but there are those few who will rush in where angels fear to tread. Indiscriminate knob twiddling, switch snapping, and button pushing should be vigorously discouraged. Not only can this lead to degradation of instrument capability, but failure of a simple switch can render an instrument unusable for days until a replacement part is obtained.

Purchase items that need frequent replacement in an instrument before they are needed and keep in stock. There is nothing more frustrating than costly downtime while waiting for a replacement part, such as a new light bulb. A little forethought and planning when a new instrument is purchased can be effective in avoiding costly shutdowns.

Modern instruments rely heavily on electricity, and the electrical environment must be optimum for proper operation. Take care that the capacity of the line to an instrument is not exceeded. Even if the line is not loaded sufficiently to blow a fuse or circuit breaker, heavy loading may cause excessive warm-up times, low voltages, or voltage surges when instruments on the line are turned on or off. One sign of excessive loading may be a microprocessor which loses memory or which cannot be programmed properly, or is not following a presumed program. This condition of near-overload can be checked by turning off the instrument, disconnecting some of the other equipment on the line, waiting a few minutes, and then turning the instrument on. If this cures the problem, then overload was the cause.

A most important consideration in modern instrumentation is proper grounding. All modern equipment is supplied with three-prong electric plugs, with the third prong connected to the case of the equipment. There are two reasons for this: safety, so that the case of the equipment cannot become "hot" in case of a short circuit, and shielding the inner electronics from electromagnetic fields caused by other equipment, for example, sparking electric motors or automotive ignition. Although all equipment today has three-prong plugs, not all laboratories have three-hole receptacles, especially older laboratories. Use is often made of an adapter for converting a two-hole socket for three-hole plugs, where the third hole is connected to the faceplate screw that connects to the box surrounding the receptacle. The assumption is that the box is grounded, but this is often not the case. The use of such adapters is not recommended. In fact the author has even found three-hole receptacles in which the third hole was not grounded. Fortunately, there are inexpensive devices that can be used to check receptacles to see if they are properly grounded. If not, an electrician should be summoned to install proper grounds. Poor grounds can cause erratic behavior of instruments, especially in establishing "baselines" or under conditions of highest sensitivity, or when other nearby instruments are turned on or off.

Another source of instrument malfunction can be voltage spikes or surges caused by nearby lightning strikes or turning on or off heavy equipment. Lightning strikes that are not close enough to burn out lines or equipment can give rise to ground currents of large magnitude, which can cause voltage spikes in the kilovolt range. The rise and fall times of these spikes may only be of microsecond duration, but this is long enough to destroy solid-state devices such as transistors, integrated circuits, and diodes. Turning the equipment off without disconnecting it will not necessarily protect it since these spikes are known to jump switch contacts, even when the switch is off.

The response time of fuses is too slow to protect against these spikes. However, there are surge protection devices, using metal-oxide varistors, or MOVs, which are relatively inexpensive and which have a resistance inversely proportional to the applied voltage and a response time of nanoseconds. MOVs can be installed in the line to which the instruments are attached or, preferably, at the service entrance line, by a qualified electrician.

Maintenance of glassware is also a quality assurance concern. Keep glassware used for trace or ultratrace analysis segregated from that used for macroanalyses. Use separate cleaning procedures for trace analysis glassware, to minimize contamination. It has been found in some laboratories, for example, that a high-temperature (400°F) bakeout for several hours may be necessary to destroy all traces of organic materials in glassware. Never use excessively scratched glassware in trace analyses, to minimize errors due to absorption on the glass surfaces.

MATERIALS

Materials used for chemical analysis are another component of the analytical system that needs control. In general most materials fall in the following categories: standards, reagents, solvents, water, filter media, and chromatographic adsorbents.

Standards. There are two classes of standards used in the laboratory: primary standards, which are those whose characteristics or composition are certified by the organization that issues and sells them, and secondary standards, which are not so certified. When the issuing organization for a primary standard is the National Institute for Technology and Standards (NITS)(formerly the National Bureau of Standards, NBS), the material is designated a Standard Reference Material or SRM.

SRMs and other primary standards are expensive materials because of the care in preparation and the work required for certifying composition, and so are rarely used for routine quality control work. However, they are useful as an occasional check on the accuracy of a method, and may be used as a reference for interlaboratory comparisons or certification of an analyst's ability to perform an analysis.

Secondary standards are those established by the laboratory for quality control evaluations, such as control charting, method validation, or checking standardization curves. They should, of course, have a purity consistent with their use and must be homogeneous.

Handle all standards with care. Never return material removed from the container to it, but discard the material if not used in the analysis. Preservation techniques, such as low-temperature storage, or storage under inert gas, may also be necessary to preserve purity or analyte concentration.

Reagents. All reagent chemicals used in the laboratory should be at least "analytical grade," as defined by American Chemical Society (ACS) specifications. Label reagents with the date received, the date first opened in the laboratory, and an expiration date, if this is applicable. In many laboratories this is done by writing the information on the label of the bottle. This is not recommended practice, since labels are prone to being torn, and information on the label may be obscured. A better technique is to use a separate label, and, in fact, labels specifically designed for this purpose are available from laboratory supply houses, supplied in tape form.

Instruct personnel to always keep reagent bottles tightly capped to avoid contamination or pickup or loss of moisture. As in the case of standards, never return material removed from a reagent bottle to the bottle, to avoid contamination. Inspect reagents stored in the laboratory periodically, at least semiannually, for signs of deterioration, and discard suspect containers. Discourage personnel from ordering larger quantities than can be normally used up in a relatively short time even if the price is more favorable.

Reagent Solutions. These are solutions prepared in the laboratory. Label them with the date of preparation, the identity and concentration of active ingredients, and an expiration date. Reagent solutions, no matter how stable, should always bear an expiration date, which should be no later than six months from the date of preparation. This is because, with constant use in the laboratory, opening and closing of the container, exposure to vapors and fumes in the laboratory, the longer the solution stands on the shelf the more likely it is to have become contaminated.

One practice which is prevalent in many laboratories and should be vigorously discouraged is pouring unused reagent solutions back into the bottle. An analyst will often pour a portion of the reagent solution into a beaker, extract an aliquot with a pipet, or fill a buret, and then pour the unused material back into the bottle. The rationale for this is conservation of costly material and time of preparation, but the possibility for contamination is obvious.

Solvents. The solvents used in the laboratory should be of a purity consistent with the analysis, that is, "spectrographic grade," "chromatographic grade," and so on. As with reagents, keep solvent bottles

Table
6-2. Grades of Reagent Water

Grade of Water	Total Matter, mg/L	Maximum Conductivity, μmho/cm
Type I	0.1	0.06
Type II	0.1	1.0
Type III	1.0	1.0
Type IV	2.0	5.0

tightly capped, and never return unused solvent to the bottle, to avoid contamination.

Many laboratories doing a high volume of chromatographic work find themselves using large volumes of expensive solvents and faced with the problem and expense of disposal of them. The idea occasionally surfaces of minimizing costs by redistilling the solvents and reusing them. In general this is not a viable option. To redistill to the requisite purity involves expensive equipment and labor, and the resulting solvent is always suspect when used for analysis.

Water. Water is the most ubiquitous chemical used in the laboratory, and hence one of the most important for control of quality. Today many treatment processes for purifying water are available to the laboratory, including distillation, deionization, membrane filtration, reverse osmosis, and UV sterilization (for microbiological work).

The treatment used must match the requirements of the analytical work being done. The American Society for Testing and Materials (ASTM)[1] has categorized laboratory-pure water in four classes or types, ranging from type I, the purest, to type IV. Table 6-2 shows the characteristics of the four different types. Achieving type I water from most tapwaters is not an easy task, requiring distilling water with a maximum conductivity of 20-μmho conductivity, followed by deionization and filtration with a 0.2-μm membrane filter. Commercial systems are available guaranteed to produce water of this type or less pure.

Use of a water softener followed by an activated charcoal filter for treatment of tapwater prior to purification is often desirable. The water softener exchanges calcium and magnesium for sodium, thus minimizing the build-up of scale in the still, and the charcoal filter removes some of the organic contaminants that may be present. Use of alkaline permanganate in the still bottom will help to destroy organics, which might otherwise be carried over, and a charcoal filter after deionization will help to remove contaminants from the ion-exchange resins used.

Storage of pure water and delivery to the laboratory can also present problems of contamination. Ideally, the water should only contact glass or tin, but this is often not convenient or practical because of cost. Analysts should be aware of the fact that solids can leach from glass and organic contaminants can leach from any plastic the water contacts. In analysis for ultratrace quantities of organics it may be necessary to extract the water used with solvent to remove organic contaminants.

Whatever treatment is used, monitor the water frequently to determine if purity has been compromised. Check conductivity at least daily, and analyze samples of pure water periodically (monthly or bimonthly) for heavy metals and organic materials.

REFERENCE

1. Water, part 31, *Book of ASTM Standards*, Philadelphia: American Society for Testing and Materials, 1977.

CHAPTER

7

Documentation for Quality Assurance

The previous chapters have been concerned mainly with the operations of quality control. Most quality control measures are essentially little more than good scientific practice, the sort of things learned in college analytical chemistry courses, but which are neglected in the hustle of getting the work out. If analytical chemistry were carried out in a vacuum, as a purely intellectual pursuit, quality control would be sufficient. However, analytical chemistry is always carried out for a purpose, that is, it is an information-generating process and its output is of real economic value to the outside world. For this reason quality control by itself is not sufficient since the outside world of clients, purchasers, accreditors, lawyers, judges, and so on, needs to be assured of the quality of the analytical data. Quality assurance arose as an answer to this problem and has been extended to assurance of the security and traceability of the data, as well as quality control. Quality assurance is an outwardly directed activity by which the laboratory assures others of the quality of the information it produces.

The four objectives of quality assurance given in Chapter 1 stress the need for *documentation* to achieve these objectives. This chapter will discuss the techniques for acquiring data, the types of data to be acquired and kept, and the storage or preservation of data.

GENERAL PRINCIPLES

There are three general principles that should apply to documentation and recording of information. These are permanence, attributability, and security.

Permanence

All information recorded should be in a permanent form. In general this means keep records in indelible ink; forbid the use of pencils in the analytical laboratory.

While this stricture seems obvious, it is surprising how many laboratories in this day and age still allow analysts to record data in pencil. Needless to say, this is completely unacceptable for quality assurance, since penciled data is too easily changed by erasure.

Enforcing this rule may sometimes be difficult, because many analysts seem to prefer using pencils to pens. When questioned as to why, they will say something about how permanent ink is, and how they prefer pencil because it can be easily changed. This is, of course, precisely why the use of pencils should be forbidden. In some cases this preference goes back to school habits where "neatness counts" and students get into the habit of scribbling data in pencil, and later transcribing it into a neater form in ink. This is a practice that should also not be permitted in the laboratory, because each transcription of data is an opportunity for error to occur.

Attributability

All information recorded in the laboratory that is worthy of being retained should be attributable to the person who recorded it. This means that such material should be dated and signed by the person making the record.

Signing should be by *full, legal signature*. This means that initials are not sufficient. The reason is simple—it is quite possible that two persons in the laboratory may have the same initials, and this could cause confusion in tracing the data that went into an analytical report. Furthermore it is much easier to forge initials than a signature.

Security

This principle requires that data be recorded, stored, and controlled in such a way that it is not easily destroyed, stolen, or tampered with.

The remainder of this chapter will be concerned with various applications of these principles to the types of records generated and recorded in the analytical laboratory.

DATA ACQUISITION

Of major concern in the laboratory is the acquisition of "primary" or "raw" data, that is, those data directly related to chemical measurements, which are used in the calculation of "secondary" data, that is, the final analytical results. In today's laboratory there are three general means of obtaining or acquiring such data: manual (handwritten), electronic (spectra, chromatograms, or hard-copy readouts from instruments), and direct computer acquisition (with instrument output directly into the computer).

Manual Data Acquisition

In most analytical laboratories today much of the primary data is still being acquired in the old-fashioned way, that is, written by hand. The completely automatic laboratory seems to still be a long way off. However, systems for recording the data vary. In some laboratories notebooks are used, while in others the system of using analytical "data sheets" is used. For good quality assurance the notebook system is preferred for the simple reason that the data sheets are too easily lost or destroyed, or tampered with.

Some laboratories use data sheets that are preprinted with blanks to be filled in with the pertinent data. This adds to the efficiency of data collection and increases productivity. One system for preserving this efficiency and avoiding loss of data sheets is to take a number of the preprinted sheets and bind them into a book. Inexpensive machines for binding (often used for research reports, manuals, and the like) are available.

Notebooks

All notebooks used in the laboratory should be bound and have consecutively numbered pages. Do not use loose-leaf binders, spiral-bound notebooks, or shorthand notebooks for the reason given above, that is, sheets are too easily torn out and lost. The numbering of pages indicates at a glance that pages have not been removed from the book.

To avoid loss or misplacement of notebooks set up a system to control distribution of notebooks. Consecutively number all notebooks on the cover and keep a file containing the number of the book, type (if more than one type or size of notebook is used), date of issue, name of person to whom it is issued, date of return, and storage location of the returned book. Responsibility for notebook control should be in the hands of one person. This may be the quality assurance officer, his or her secretary, or another designated person. Using this system ensures that the location of each notebook containing valuable data is known at all times.

Notebooks do not have to be returned as soon as they are filled. Many analysts need to keep filled notebooks for a period of a few months after they are completed, for reference, but analysts should know that they are personally responsible for the notebooks until they are returned to the issuing office for storage.

In general it is good practice from a quality assurance standpoint for each analyst to have his or her personal notebook. While this is feasible in many laboratories, in others it is less desirable from the standpoint of efficiency, especially in those laboratories where large numbers of samples are run with a division of labor (e.g., one person weighs samples, another prepares them, and a third makes a measurement of some kind). Here there is often one notebook for each operation, which is used by the person performing that operation, who may not be the same person every day. While this type of operation is acceptable, provided each user signs and dates his or her entries, the notebook itself should be issued to an individual, for example, the laboratory supervisor or other reliable employee.

Data Entry

Instruct all employees in the proper way data should be entered in a notebook.

- Enter data directly into the book as it is acquired. It is not acceptable to record data in pencil on a scrap of paper and then later enter it into a book. This practice can lead to errors in transcribing the data.
- All data entries should be in nonerasable ink. Allow no pencils to be used.
- All data entries should be signed by the person making the entry and dated. Use a full legal signature (no initials). If the entry takes up more than one page in the book, separately sign and date each page.
- Cancel errors by drawing a line through the entry in such a way that the erroneous entry can still be read. Do not allow the use of ink eradicator or correction fluids, for example, "liquid paper" or "white-out." It is recommended that a note be made in the margin indicating the reason for the cancellation.
- Mark blank pages or substantial portions of pages with no entries with a large X to indicate that they were intentionally or inadvertently left blank and cannot be used for additional data entry at a future date.
- In addition to the signatures of those making the data entries, the laboratory supervisor or his or her assistant should also sign the book periodically, for example, every 25 pages. The signature should follow a statement such as "read and understood by —." A rubber stamp with the phrase may be used, but the signature should be genuine. The purpose of this rule is to demonstrate that the supervisor or other responsible person is aware of the way the notebook is being kept.

Electronic Data Acquisition

By electronic data is meant the output of various instruments or the electronic circuitry associated with the instruments. Examples would be GC or HPLC chromatograms, UV/visible or infrared spectra, or hard-copy printouts such as are obtained from some modern atomic absorption spectrometers. The following are some rules that should be followed for good quality assurance:

- All such printouts, chromatograms, spectra, and so on, should be signed and dated by the analyst responsible for the analysis. Some modern instruments feature a convenient printout of the analyst's name at the push of a button. This is not acceptable as a replacement for the signature, since anyone can push the button. A signature is difficult to forge.
- All such printouts should contain the laboratory ID number of the sample. In some cases numbers assigned by the analyst (1, 2, 3, and so on) are used to identify printouts. This is not acceptable since it may make traceability difficult.
- Chromatograms and spectra usually contain multiple peaks, only one of which corresponds to the analyte of interest. Clearly identify this peak on the printout so that at a later date there is no ambiguity regarding which peak was used.
- Although the electronic printouts cannot be kept in a bound notebook, file them in a secure place, using a system that permits easy retrieval.

Direct Computer Data Acquisition

Direct computer data acquisition occurs when the output of an instrument is channeled directly to the input of a computer, without human intervention. Although this has long been recognized as a desirable use for computers, there has not been widespread application in analytical laboratories for a number of reasons. One reason is that the low price of microprocessor chips has led to the incorporation of the computer directly into the instrument. The result is that the analyst sees only the calculated output from the combined instrument/microcomputer. These data are then usually manually recorded in a notebook, or entered into a mainframe computer for input to the laboratory information management system (LIMS) for calculation.

Another concept of more recent vintage is the "electronic notebook," which is being vigorously pursued in various laboratories. Originally the electronic notebook was visualized as a small, battery-operated computer that could be easily transported. Its use was to be limited to collecting data from instruments such as balances and automatic titrators, which could then be dumped into a mainframe or minicomputer for further calculation. However, the development in recent years of inexpensive, powerful personal computers has made it possible to place individual computers at many of these instruments for data collection and manipulation.[1]

The following are some basic principles related to direct computer data acquisition for quality assurance purposes:

- "Validate," that is, check for accuracy software used in the process. Probably the best way to do this is to compare the computer-generated results with hand-calculated data. Do this on a variety of data, to ensure that there are no software "bugs" that would show up only on certain selected data. This type of validation procedure should also be run at a time when the computer is being used at close to maximum capacity, to ensure that bugs do not exist under these circumstances.

  Thoroughly document in-house written programs with the signature of the programmer, date, and revision number.
- Reduce computer-collected data to hard copy as soon as possible. Archive the hard copy. Do not rely on computer disks or tapes for long-term data storage.
- During data collection, use disk storage and backup disks frequently to avoid problems such as power outages or "head crashes," which can destroy data and programs.

ANALYTICAL METHODS

In Chapter 5 we discussed analytical methods and the format in which to write them, as well as the types of methods and ways to evaluate them. Keep all authorized analytical methods in a methods manual in the laboratory, where they are accessible to all analysts.

In many laboratories the methods manual is kept in the supervisor's office. This means that the manual may not always be accessible to all analysts, since there may be times when the office is locked or the supervisor is consulting with other persons, and the analyst may be hesitant to interrupt. For these reasons keep the manual in the laboratory proper, where any analyst who needs to refer to it may do so.

The methods manual should be a loose-leaf type so that methods can be removed when they are no longer current. It is important that the manual only contain methods that are currently in use, to avoid having analysts use the wrong method. On the other hand, do not discard obsolete methods, but rather relegate them to the historical file, which may be kept by the supervisor or the quality assurance officer.

In Chapter 5 the importance of numbering methods was stressed so that there is never any doubt regarding which method was used for a given analysis. However, it is not necessary to issue a new number for a minor modification of a method. The modified method may be given a revision number. Note the date of the revision on the method, as well as the revision number. This should only be used for minor modifications of a method. A major change in a method should get a new number, after the validation study, and after being authorized.

Record data obtained during a validation study of a method in a laboratory notebook similar to those used for recording primary or raw data. Make data entries in the same way. To avoid confusion, use a separate notebook for method validation studies, so that raw data and validation data are not mixed up.

SAMPLING AND SAMPLE HANDLING

In Chapter 4 sampling and the requirements for documentation with regard to quality assurance were discussed. As discussed, prepare a written sampling plan before samples are taken, and write out sampling methods.

Sampling Methods Manual

As in the case of analytical methods, collect sampling methods in a manual, which is kept in the laboratory, accessible to analysts and samplers. Only current methods should be kept in the manual, which should preferably be of the loose-leaf type. Keep obsolete methods in a historical file, for traceability purposes.

If sampling is conducted at a remote location, take steps to ensure that samplers have available copies of the sampling method to be used. This may consist of leaving a copy of the method at the sampling site, keeping a copy in the vehicle used to transport samplers, or requiring samplers to carry a copy with them.

Also number sampling methods. The number can be indicated in the log-in record of the sample, thus documenting the procedure used to obtain the sample.

Sample Labeling

As soon as a sample is taken, prepare a label and attach it to the sample container. Various systems are used, depending on the circumstances, and the number of samples taken. For example, if a large number of samples are taken, the containers may simply be numbered, and a sample log sheet or book used to enter the necessary information opposite the sample number. Sample labels should be of such a type that they are not easily destroyed or separated from the container. For example, labels on aqueous samples should be waterproof. Use nonerasable ink to record sample information. Do not use crayon pencil to number samples since it is easily rubbed off. One important point here that is obvious, yet frequently overlooked, is that the label should never be attached to the lid or cap of the container, but only to the container itself. Lids and containers can easily be interchanged, leading to mislabeling of samples.

The minimum information required on a sample label, or sample log sheet, is

Sample designation: a number or alphanumeric symbol unique to the sample.

Description: a brief description of the type of material sampled, for example, "wastewater," "reactor bottoms," "soil."

Location: an unambiguous description of the place where the sample was taken.

Time and date of sampling.

Sampler: full name of person who took sample.

Sampling method number.

Additional information, such as temperature, or the results of spot tests performed at the time of sampling (pH, chlorine residual, and so on) may also be included if these are important to the analysis or interpretation of results.

Sample Log-In

Immediately on receipt at the laboratory the samples should be "logged-in" by a responsible person. Sample log-in is a task that should be reserved to certain designated persons. In other words, the practice, which is common in some laboratories, of allowing anyone to log-in samples should be discouraged. Only in this way can management be sure that the procedure is carried out correctly. Sample log-in is an important step for legal purposes to document the fact that the samples were received in the laboratory. It can also be important in tracing a sample that may have been lost.

In some laboratories sample log-in is done manually, using a bound, ledger-type book, using the same techniques for entering the data as described above for laboratory notebooks. In many laboratories today, computer log-in is common. All pertinent data are entered into the computer, which assigns the laboratory identification number and in some cases automatically prints out a label for the sample, including perhaps a bar code for automatic readout in the laboratory. The database thus created can be used as the basis for a sample management system that tracks the sample through the laboratory, does the calculations, and generates the final report.

However, if computer log-in is used, frequent backup of the data is recommended and hard-copy printout at least daily is necessary. Do not use the computer or computer disks for long-term storage of data. The hard-copy printouts can be stored as documentation of the receipt of the samples. The sample log-in contains the following information:

- *The laboratory-assigned sample number.* Various systems are used for numbering samples, using numbers or alphanumeric designations. For example, a letter may be used to designate the matrix, such as F for food, W for water, followed by a consecutively assigned number, or each sample may simply be given a consecutive number. The important point here is that the number be unambiguous, that is, no two samples shall be given the same number. Although this seems obvious it is not always practiced, especially in laboratories where different departments are allowed to log-in samples independently of each other.
- *Name of the client, that is, the person requesting the analysis.* In some cases it may be important to also give the name of the client's organization or company, if different from that of the laboratory.
- *Client's sample designation, if appropriate.* This is the number or description used by the client to describe the sample.
- *Sampler's name, if known.* If the sample was taken by the client, this should be indicated.
- *Sampling method number.*
- *Nature of the matrix.* Water, soil, organic liquid, solid waste, and so on.
- *Analyses requested.*
- *Sample storage location.* Most samples are not analyzed immediately upon receipt at the laboratory. They must be stored somewhere until they can be integrated into the laboratory operation. Set aside and designate sample storage areas to guard against sample loss and/or mix-up.
- *Special storage conditions.* Samples may require refrigeration, storage under inert atmosphere, in the dark, and so on. Enter this information on the sample log-in record. It may also be of advantage to log an expiration date, after which the sample is to be discarded after analysis. This will help to prevent the accumulation of large numbers of samples that are no longer of interest.

Sample log-in then provides documentation that

1. The sample was, indeed, received in the laboratory on a certain date.

2. It was sampled using a known sampling method.
3. It was assigned a unique identification number.
4. It was stored in an identified location until analyzed.
5. It was preserved using an approved method so that, presumably, it was in the same physical and chemical state when analyzed as when collected.

Forensic Samples

Forensic samples are those whose analyses may become the basis of a dispute, such as a legal case, a case of compliance to a government regulation, or a dispute between a buyer and seller. In these cases the legal concept of "chain of custody" may be important. Chain of custody requires that the sample be in the custody of some known person from the time it is collected until it is disposed of. A sample is said to be in custody if it is in the person's hands, within eyesight of the person, or in a locked storage space with limited access.

Chain of custody procedures thus require

1. Receipt of the sample in the laboratory is documented, with signatures of the person submitting the sample and the person who accepted it.
2. Storage of the sample is documented, with the signature of the person who put it in storage.
3. A locked storage area is used, with limited access (i.e., only one or two persons have a key).
4. Analysts "sign out" the sample when removing it from storage and keep track of it in the laboratory. If the analysis takes more than one day, the sample is returned to storage and removed again the following day, and these steps are documented.
5. Finally, in some cases disposal of the sample is documented.

In most laboratories this chain of custody procedure is not necessary for the majority of samples. However, if a sample is to be analyzed for forensic purposes, or even if it is possible that the results may figure in a dispute, the chain of custody procedure is used.

INSTRUMENT CALIBRATION AND MAINTENANCE

For traceability purposes document all calibration runs and maintenance work done on instruments and other equipment. If an analytical result is questioned, the fact that a major component of an instrument showed signs of malfunction or needed to be replaced on or about the time the analysis was run could be an important clue to the cause of a poor result. Conversely, documentation that the instrument was calibrated at that time, with no sign of malfunction, can eliminate the instrument as a source of error.

Keep calibration and maintenance records in the same manner as primary data, that is, in bound notebooks, in ink, and suitably dated and signed. Record all results of calibration work. There is a temptation not to record calibration results if no problem was found, but it is important to note the fact that the calibration was run and no untoward results found.

Maintenance work is frequently done by outside personnel such as an instrument technician, or manufacturer's representative. In these cases, the laboratory person responsible for checking the instrument after repair should document that the work was done, and by whom, and when. Instruct personnel that it is important that *all* maintenance work be documented, no matter how minor it may seem.

Keep calibration and maintenance notebooks in the laboratory, near the instrument, to ensure that the work is recorded. Each instrument should have its own notebook, for ease of reference, although the same notebook may be used for both calibration and maintenance.

Calibration and Maintenance Methods

Write down procedures for calibration and keep in a calibration methods notebook, which may be of the loose-leaf type similar to the analytical methods and sampling methods manuals. In many cases calibration procedures are given in the manufacturer's instrument manual. These need not be transcribed into the calibration manual, but should be referenced in the manual, or better yet, a photocopy may be placed in the manual. Instruct personnel to only use the prescribed calibration procedures, unless special permission is given to use alternative methods.

Most maintenance work on instruments and equipment is performed on an as-needed basis, that is, when the instrument fails or malfunctions. However, some equipment requires periodic preventive maintenance, and written procedures should be prepared for this and kept in a manual similar to the others described above. It may also be of advantage in some cases to prepare a "diagnostic" or troubleshooting manual for complex instruments to indicate the maintenance needed to correct certain malfunction problems. This should include those corrective actions taken in the past when the instrument exhibited certain problems, and could be an important factor in minimizing downtime.

REAGENTS AND REAGENT SOLUTIONS

The purity of reagents and the composition of solutions prepared from reagents are of paramount importance in obtaining accurate analyses. One of the major sources of blunders, that is, outliers, in analytical results is contaminated reagents or solutions that have been incorrectly prepared. Preventing these errors, or at least providing a means of identifying them, is one of the objectives of a good quality assurance program.

Purchasing reagents with the necessary degree of purity is well understood by analytical chemists, and manufacturers have done an admirable job of providing reagents of varying grades to suit the needs of chemists. The American Chemical Society, the U.S. Pharmacopoeia, and other organizations have assisted by setting standards that permit manufacturers to label reagents: ACS Grade, U.S.P., Analytical Reagent, and so on, with labels containing the maximum contamination limits. Although mislabeling of reagents can occur, it is so rare that it is seldom even considered as a possible cause of erroneous results.

The problems with reagents arise after they are received in the laboratory. As soon as the container is opened, the possibility of contamination is present. Bottles that are not tightly closed expose the contents to the air with possible loss or pickup of moisture, absorption of carbon dioxide, or absorption of contaminating vapors that may be present. Material may be removed with a contaminated spatula, especially by nondegreed personnel who often do not understand the extreme care necessary to prevent contamination.

Always label purchased reagents with the date received and the date the bottle was opened. In addition an expiration date may be added if the material is likely to degrade over a period of time. While such labeling will not prevent contamination problems, it will alert the experienced analyst to the possibility of contamination or degradation at the time of analysis, or when a suspicious result has been obtained.

Use separate labels for this purpose. Do not write the relevant dates on the existing manufacturer's label because such inscriptions are often difficult to read, or may obscure important information on the label. Suitable labels are commercially available, printed with the appropriate legend, with blank spaces for the insertion of dates. These may be affixed to the bottles when received and inscribed as necessary.

Label reagent solutions prepared in the laboratory with the date of preparation, concentrations of active ingredients, and an expiration date. Even solutions that are known to be stable should be given an expiration date, beyond which they should not be used. The reason is that solutions are generally used frequently, opened and closed again, with more or less airtight seals. The possibility of contamination, evaporation of solvent, and leaching of impurities from containers increases with time. Do not use any reagent solution more than six months old in an analytical laboratory.

One practice which is prevalent in many laboratories, and which should be vigorously discouraged, is pouring unused solutions back into the bottles. An analyst will often pour a portion of the reagent solution into a beaker, extract an aliquot with a pipet, or fill a buret, and then pour the unused material back into the bottle. The rationale for this is conservation of costly material and time of preparation, but the possibility for contamination is obvious.

Every laboratory should have a Reagent Preparation Notebook. This is a bound notebook, kept as previously described, that is, entries in ink, signed and dated, and so on. When a reagent solution is prepared, the analyst records the date, the actual quantities of weight and volume used, and his or her signature in the notebook. This will be of great assistance in tracing possible sources of error in analyses.

Analysts, being human, are quite capable of preparing a 10-percent solution when a 1-percent solution is desired, for example, or of using sodium carbonate instead of bicarbonate. If forced to record actual quantities and materials used, such errors are easily detected when tracing possible sources of error.

PERSONNEL TRAINING RECORDS

Clients, or users of analytical laboratory data, assume that the data are the result of work by analysts thoroughly trained in laboratory operations. Thus, the question of adequate training may arise in cases of dispute over the quality of the results. It is good management practice, therefore, to maintain files documenting the training of personnel.

Professional (Degreed) Personnel

In general the analyst with a four-year college degree with a major in chemistry may be considered to have learned the basic laboratory techniques necessary to function as an analyst with minimum on-the-job training. He or she may not have had extensive training in modern instrumental analysis, but will have been exposed to the basic concepts of chromatography, absorption and emission spectrophotometry, and electrochemistry, and should be able to master instrumental analysis using these techniques with a minimum of training and supervision.

Professional personnel, on the other hand, often turn to continuing education courses in special topics in chemistry or related topics such as electronics, data processing, or laboratory management (such as quality assurance). To the extent that such courses relate to the work of the laboratory record them in the individual's personnel file along with pertinent information such as the institution offering the course and the date completed. Attendance at minicourses, seminars, and the like, which may be given by instrument manufacturers and professional organizations, should also be recorded in the personnel file.

These records not only serve as documentation of relevant skills, but are valuable reminders of the person's value to the organization at the time of the annual performance review.

Technician (Nondegreed) Personnel

Newly hired technicians, unless they have previously been employed in a similar laboratory, are usually given on-the-job training by a supervisor, another professional, or an experienced technician. It is important that the extent of their training be documented, particularly as to the dates they were judged proficient enough to perform certain analyses. The dates are important because this will establish the extent of their experience at some future date, when this might be questioned. The documentation may relate to the mastery of common laboratory skills such as gravimetry, titrimetry, and so on or may simply indicate the date they were deemed capable of running a given analytical method. In the latter case, record any hard data to prove this, such as results obtained on a standard material of known composition, or comparison of results obtained by the technician with those obtained by an experienced analyst.

As in the case of professional employees, keep records of relevant courses taken, seminars attended, and so on. Videotape training of technicians in selected topics is becoming widespread, and exposure of a given technician to such tapes should be recorded, along with the date and results of any test that might have been given.

Responsibility for Personnel Training Records

Personnel training records for both professionals and technicians should be kept by the laboratory supervisor, since he or she is responsible for the training of the analysts. Keep them in the supervisor's office, for convenience in updating. Copies may be made for inclusion in the employee's regular personnel file, which is usually kept by the Personnel Department.

ANALYTICAL REPORTS

The analytical report is the final product of the analytical laboratory and is often the only contact the lab has with its clients. It is important that the report be complete, accurate, and easily understood by the person who requested the analysis. The following guidelines will assist in achieving these objectives:

1. Include the following information on the top of the first page, or introduction to the report:
   - **(a)** The name of the laboratory that performed the analyses. Address and phone number may also be included.
   - **(b)** The name and organization (if necessary) of the client. Never only the name of the organization without the contact person. Reports go to individuals, not organizations.
   - **(c)** The date the sample was submitted, and the date the report was issued. In some cases it may be important to give the date and time the sample was taken.
   - **(d)** The name of the person who took the sample. If the sample was taken by the client's personnel, note this in the report.
2. Include in the body of the report the results of the analyses that were run and the following information:
   - **(a)** The laboratory-assigned sample number.
   - **(b)** The client's sample number, or other designation.
   - **(c)** The name or appropriate designation of the analytes, that is, the compounds, elements or groups of compounds for which the analyses were run. Take care here to avoid ambiguities or possible misconceptions on the part of the client. Use chemical names wherever possible, but common names may be used, if there is no chance of their being misconstrued. For example, the name poly-(1,4 anhydroglucose) in place of the common name "cellulose" could lead to confusion. On the other hand, always avoid abbreviations. The analyst may know that PCB stands for polychlorinated biphenyls, but there is no guarantee that the client does.
   - **(d)** The actual numerical result of the analysis. Take care here.

Never report a result of "zero" since no chemist can ever be sure that there is zero concentration of anything in any sample. Instead use the "less than" sign (<), followed by the detection limit of the analysis, if known. If the detection limit is not known, the phrase "none detected" should be used.

All analysts should be aware of the proper reporting of numerical results with respect to significant figures, that is, the digit furthest to the right represents the only uncertain digit in the result. Train nondegreed analysts (technicians) in the principles of reporting scientific data, or have a chemist carefully scrutinize their reports. In particular they are prone not to report trailing zeroes after a decimal point when the zeroes may be significant or else go to the opposite extreme and report all the digits their electronic calculator can provide, regardless of significance.

(e) Give some thought to the reporting of units of concentration. In general, avoid dimensionless units since they do not indicate whether the concentration is weight by weight, weight by volume, or volume by volume. Thus mg/L is preferred to ppm because the latter does not indicate weight by volume. In dilute aqueous solutions there is no difference because the density of water is 1, for all practical purposes, but in reporting the concentration of a solid contaminant in air, for example, there can be a large difference. In the latter case mg/M^3 would be a better choice of units. Concentrated solutions, often reported in percent, should be clearly designated as g/100 g, g/100 mL, or % (by weight).

Some clients may prefer to have results reported in units not customarily used by chemists, for example, g/ton or lb/gal. In such cases it is wise to accommodate the client, since the analyst can be sure the conversion is done correctly, if he does it himself.

(f) Assign a code number to each result, which designates the analytical method used and the analyst who ran the analysis. This will be very useful when tracing questioned data in the future.

(g) The actual format of the report will depend on the nature of the report. For example, if a number of analyses were run on a single sample the report may cover only that sample, listing analytes, data, units, and analysis code. On the other hand a report may cover a series of samples, which were analyzed for a single analyte. In this case the report will list sample numbers, analyte concentrations, units, and analysis code.

3. Show the name, title, and signature of the laboratory supervisor, or his or her designated representative, on the bottom of the report. This is a vital part of the report since it documents that the signatory has reviewed the results, approves them, and accepts responsibility for their quality.

4. If a report consists of more than one page, number each page, using a format such as "page *x* of *y*" where *x* is the page number and *y* is the total number of pages. This enables the client to tell immediately whether a page is missing or not.

FILING QUALITY ASSURANCE DOCUMENTATION

A good quality assurance program generates a great deal of paper records as the reader of this chapter will have guessed by now. To obtain the maximum benefit from this paper establish a system for rapid and easy retrieval of any given document. Quality assurance is undertaken for the benefit of outsiders, for example, clients, government agencies, and accrediting organizations, and they will want to inspect the documentation of the system. In many laboratories the documents are not kept in a centralized location and little attempt is made at document control. This can lead to a great deal of unnecessary confusion during inspection, and unnecessary embarrassment for the laboratory manager.

The answer to the problem is a good filing system. Keep all documents involved in quality assurance, or suitable photocopies of them, in a central file, preferably in the quality assurance officer's office, or where he or she will have ready access. Tailor the actual system used to the individual laboratory. A good secretary will be invaluable in helping to define the system. The author claims no special expertise, but the following is offered as a suggested system.

First it is necessary to decide which documents are going to be filed and to divide them up into relevant main groups. Each group is then given a title and a two-digit number. Within each group a series of file folders is set up with a title and a number consisting of the main group number and a second two-digit number unique to that file folder. For example, the main group number 02 may stand for Performance Evaluation Reports, while a folder containing results reported to the EPA may have the number 02-01 and the title "EPA Drinking Water." A master index is prepared, listing the various file folders by name and number, and each document filed has the appropriate number assigned to it prior to filing. In this way improper filing is minimized. It is also good practice to limit the persons permitted to file documents, for example, the quality assurance officer and his or her secretary.

While each laboratory must determine its own filing system, the following is a possible list of main group titles:

Quality control charts. Keep these in the laboratory, and forward copies to the quality assurance officer to be filed.

Performance evaluation test results. File by name of the organization that issued the sample and by type of sample.

Standard reference materials. Copies of certification documents and test results.

Blind-sample programs. File by sample type (matrix).

Accreditation–inspection reports and correspondence. File by name of accrediting organization.

Accreditation certificates. File by accrediting organization.

Quality assurance irregularities/deficiencies. File chronologically.

Quality assurance standard operating procedures. File by number of the SOP.

Quality assurance audit reports. File chronologically.

Copies of analytical methods for the quality assurance officer's historical file may be kept in numerical order in loose-leaf binders in the quality assurance officer's office. Calibration data and equipment maintenance data are best kept in the laboratory for reasons previously given. Personnel training data should be kept by the laboratory supervisors with copies in the employee's personnel file, and updated annually.

A good filing system will greatly increase the laboratory's efficiency and ability to respond to requests for information regarding quality assurance. It will also ensure that the relevant data will be retained.

REFERENCE

1. Schumaker, David. Paper delivered at Association of Official Analytical Chemists, annual meeting, St. Louis, MO, September 28, 1989.

CHAPTER

8

Organizing for Quality Assurance

The previous chapters have discussed most of the necessary elements of a good quality assurance program. This chapter is concerned with the planning and organizing that must be done to establish such a program in an existing laboratory, which may have some of the elements but not all. Many laboratories have various aspects of quality control in place but are deficient in quality assurance. Quality control is generally nothing but good scientific practice, but the elements of documentation, sample control, traceability, and so on, are often woefully lacking in many laboratories.

MANAGEMENT COMMITMENT

To establish a good quality assurance program it is absolutely essential that senior or top management in the laboratory be totally committed to this goal. Management must also communicate this commitment forcefully to the laboratory staff, so that there is no doubt in employees' minds that this is what is expected of them.

Unfortunately, in many laboratories this commitment is either lacking or half-hearted. One reason is the perceived or imagined cost of the quality assurance program. In Chapter 1 we discussed the reasons why management should be committed to this aspect of laboratory management. Although the cost of quality assurance can be easily measured with standard accounting techniques, the benefits are not so easily demonstrated. In fact the quality assurance program should be perceived as an element of the laboratory's "risk management" or "loss prevention" program, in addition to the safety program, security costs, or insurance costs.

One thing is certain. If management is not totally committed to the program, this will quickly be perceived by the laboratory staff, even if management pays lip service to the concept. The result will be loss of rigor in applications of the program and conscious or subconscious sabotage of the program.

THE QUALITY ASSURANCE OFFICER

Since managers are not in close contact with the operations in the laboratory on a daily basis, it is necessary to delegate responsibility for monitoring and evaluating the quality assurance program. This is customarily done by the appointment of a quality assurance officer. Some laboratories may seek to achieve the same effect by establishing a quality assurance committee, but this is not recommended. Committees generally lack the sense of responsibility needed to get the job done, and they tend to be inefficient. Committee members perceive the job as an unwanted intrusion on their normal working duties and will procrastinate and fail to give the job the attention it requires. If the committee has a strong chairperson who can steer it successfully, this will help, but in this case, why not appoint the chairperson quality assurance officer?

Quality assurance officers have a variety of titles in the laboratory community: Quality Assurance Director, Manager, Supervisor, Coordinator, and so on. The choice of title depends on personal preference or the system used in the laboratory or the parent organization. In manufacturing operations, where product quality control and quality assurance are practiced, the title may be preceded by the word "Laboratory" as in "Laboratory Quality Assurance Manager," to emphasize the fact that the responsibilities are concerned with laboratory quality assurance and not product quality assurance.

In large laboratories it may be necessary to have a staff of several persons engaged in quality assurance, while in a small laboratory there may not be enough work to justify a full-time quality assurance officer. The size of the quality assurance staff will depend on both the volume of samples analyzed and the variety of analyses run. A large laboratory running a large number of similar samples for a limited number of analyses may well require less quality assurance documentation, control charts, data, and so on, than a small laboratory, which may be running a broad spectrum of analyses on a variety of types of samples.

Another factor impacting on the amount of quality assurance work is the perceived need for quality assurance. A crime laboratory, since most of their work will be scrutinized in a court of law, will require more quality assurance than a small control laboratory in a manufacturing operation. Most commercial testing laboratories have strong quality assurance programs to suit their clients' needs, and because they are an excellent marketing tool to interest prospective clients.

Carefully consider the choice of a suitable person for quality assurance officer, since he or she will be the key person in the program. The following are some guidelines for choosing a quality assurance officer.

Education and Experience

The person chosen should have a minimum of a bachelor's degree in chemistry, with some years of experience at the bench level in an analytical laboratory. This is an essential requirement in spite of the fact that many laboratories do not adhere to this standard. The quality assurance officer must interact with the laboratory's clients and technical personnel and must be able to call on a good chemical background to understand both constituencies. While biologists, chemical engineers, medical technicians, chemical technicians, and so on, may have had some training in chemistry, they do not have the depth of knowledge necessary to function effectively without calling for chemical assistance, which weakens their position.

It is also necessary that in addition to the bachelor's degree the quality assurance officer have experience in analytical chemistry, preferably in the laboratory he or she will serve. A chemist whose experience is in synthesizing organic compounds cannot be expected to acquire thought habits or mind-set of an analytical chemist very easily. The more the quality assurance officer understands about the operations in the laboratory, the more effective he or she will be in establishing the quality assurance program and resolving problems of technical personnel.

Knowledge of Statistics

Familiarity with the statistical evaluation of data is essential to the successful quality assurance officer. While an in-depth knowledge of all aspects of statistics is not required he or she should be comfortable with the uses of statistics and the interpretation of statistical tests. Although the mathematics of statistics is relatively simple, especially in these days of computers, the concepts of statistics and its parent field of probability can be rather subtle. Some of this can be obtained by reading the analytical chemistry literature, but it is recommended that the quality assurance officer should have been exposed to at least one good course in mathematical statistics.

General Characteristics

The quality assurance officer should be a person of maturity, and be able to interact effectively with other persons. The duties of the position will require contact with laboratory personnel at all levels as well as with outside persons representing clients, accrediting agencies, government personnel, and so on. In dealing with laboratory personnel he or she must represent the client and the client's concerns with quality. On the other hand he or she must be prepared to discuss and demonstrate the quality assurance program to outsiders. Both aspects of the job may lead to conflict and the quality assurance officer must be able to handle it with equanimity and sensitivity to the other person's viewpoint, without compromising on quality. In modern management jargon, the quality assurance officer must be "assertive" but not "aggressive." Since he or she will often be required to persuade people to do things differently than in the past, or even insist that they do so, he or she must become accustomed to criticism, often unjust criticism. This means that the job is not one for the "thin-skinned."

Since a good part of the job will require writing reports and letters, communication skills, both written and oral, are required. It will also be desirable if he or she were "computer literate," at least to the extent of being able to communicate with a computer and to understand elementary computer programming.

Duties and Responsibilities

The quality assurance officer should be the focus of the quality assurance program and the resident expert in the laboratory on all matters concerning quality assurance. Some of the duties and responsibilities of the quality assurance officer are

- Formulating and documenting procedures to achieve the quality assurance objectives.
- Consulting with the laboratory's clients on the quality assurance program and the client's needs.
- Monitoring and maintaining records of data generated in support of the quality assurance program.
- Conducting audits to assess compliance with the quality assurance program.
- Investigating deficiencies and irregularities in quality assurance and reporting them to senior management.
- Consulting with and advising technical personnel on quality assurance procedures, data, method evaluation, and so on.
- Writing and maintaining the quality assurance manual.
- Coordinating and supervising on-site inspections by accrediting agencies.
- Corresponding with accrediting agencies regarding accreditation.
- Monitoring proficiency sample testing.

- Supervising, coordinating, and evaluating blind-sample programs.
- Maintaining files on quality assurance data and correspondence.

Place in the Organization

There are two important criteria involved in placing the quality assurance officer in the laboratory's table of organization. The first is that the officer must report directly to senior or top management, and the second is that he or she must not have direct responsibility for the generation of the analytical data.

The quality assurance officer should either be a member of senior management or be solely responsible to senior management. He or she should receive authority and responsibility directly from the highest level of management in the laboratory. In no circumstance should he or she report to the laboratory supervisor, technical director, or other person responsible for the productivity of the laboratory. The reason for this is that the quality assurance officer must be perceived by laboratory personnel and outside observers as an independent monitor of the quality of the data being generated by the laboratory. There should be no real or perceived conflict of interest.

This requirement of independence is difficult to satisfy in many laboratories. If the laboratory is not large enough to require a full-time quality assurance officer, it is unlikely to have available a person with the required technical background who is not also involved in generating analytical data. The following are some suggestions for solving this problem.

If the laboratory is part of a large organization employing a large number of chemists, it may be possible to appoint a chemist not working in the analytical laboratory to the position of quality assurance officer, with the understanding that he or she will devote a certain percentage of time to the task. Such a person might be borrowed from the R&D department, plant technical department, or customer service department, for example. Needless to say this person must work in an area geographically close to the laboratory. It is virtually impossible to perform effectively as quality assurance officer from a remote location by mail or telephone.

Large organizations often have more than one analytical laboratory located at the same site. In this case an ideal solution is to exchange personnel as quality assurance officers. A chemist from lab A may serve as quality assurance officer for lab B, and vice versa. Such an arrangement can offer excellent training for the chemists involved, will improve communications between the laboratories, and may improve laboratory operations by cross-fertilization of ideas.

A variation on this can be applied in the case of a single laboratory, provided it is large enough to be divided into two or more departments. An individual can be appointed quality assurance officer for all departments except the one in which he or she works, and a person from one of the other departments appointed to monitor quality assurance in the laboratory of the other person's prime technical responsibility.

Another possibility for medium-sized laboratories that require a part-time quality assurance officer is to combine the job with another staff position that also does not require a full-time person. Many laboratories combine the quality assurance function with the safety officer position.

A final possibility that may be explored is to hire a part-time quality assurance officer. Many communities contain retired chemists, housewives and mothers with analytical chemistry training, or college professors and high school chemistry teachers who may be available for part-time work in the laboratory. A retiree from the laboratory itself would be an ideal choice for the position, as he or she is already thoroughly familiar with the work of the laboratory.

RESPONSIBILITY FOR QUALITY ASSURANCE

A good quality assurance program requires participation by all levels of employees in the laboratory. Similar to a good safety program, all employees must have their consciousness raised regarding the objectives of the program and must be aware of their responsibilities to see that the program works.

Senior Management

As discussed previously, senior or top management in the laboratory must be committed to the establishment of the program and must make this commitment clear to all employees of the laboratory. Initially this may be done by oral or written communications to the employees, but these communications must be backed up by action.

One of the most important responsibilities of management is the appointment of the quality assurance officer and the delegation to him or her of the necessary authority to get the program started and to continue the program once it is begun. There should be no doubt in employees' minds that the quality assurance officer speaks with senior management's voice in his or her discharge of duties.

Senior management's commitment to quality assurance extends to authorizing the expenditure of the necessary funds to get the job done. Although this may involve the purchase of equipment, the most important factor is the allocation of manpower. A quality assurance program such as is described in this book will involve a considerable expenditure of manpower and, if management is hesitant in allocating this, the program will not be implemented. In addition to the need for a quality assurance officer, the analysis of check standards and duplicates for control charts, materials such as notebooks, establishment of archives, and so on, there will be the need for training sessions for the employees, attendance at meetings devoted to quality assurance, and other employee-related expenses. It is quite possible that an increase in staffing may be needed. If management is seen to be excessively parsimonious in allocating finances to quality assurance the message will quickly be transmitted to employees that management wants to talk a good program, but will not support it, and this will quickly sabotage the work.

Management must also play its role in rewarding or penalizing employees for their attention to and acceptance of the quality assurance program. Include an evaluation of the attitude toward quality assurance in the annual performance review of employees.

Supervisory Personnel

In most laboratories supervisors are effectively "working managers," that is, while spending part of their time supervising other employees, they also participate in the work of their department. In their role as supervisors they are responsible to see that the persons working under them comply with the quality assurance rules and operations that form part of the program.

A major responsibility of the supervisor is to train people in the proper techniques for ensuring that quality data are produced in their laboratory. Keeping personnel training records is part of the job, as well as overseeing the use of notebooks, evaluation of control charts, archiving data and reports, labeling of reagent solutions, timely calibration of equipment, and maintenance of equipment. In other words all of what might be called the "nuts and bolts" of the program are the responsibility of the laboratory supervisor.

Another supervisory responsibility is to keep the quality assurance officer apprised of all matters in the laboratory that can impact on data quality. The quality assurance officer should be present at all discussions with clients on proposed work so that he or she can have input into the plans and ensure that the client's quality needs are met. Encourage supervisors, who are on the front line so to speak of where the work is being done, to take a positive attitude toward quality assurance and make constructive suggestions wherever possible.

Quality Assurance Officer

Although the duties and responsibilities of the quality assurance officer were discussed above, some general principles can be discussed here. Note that the quality assurance officer is not responsible for enforcing the quality assurance program. To continue the military analogy, he or she has a staff position, not a line position. The job of the quality assurance officer is to plan the program, monitor and evaluate the program, and bring to management's attention any irregularities or deficiencies discovered. The job of implementing the program and enforcing the rules and operating procedures belongs to the supervisors and, ultimately, senior management.

Do not take lightly the responsibility to keep management informed of any problems with the quality system. One of the maxims of management philosophy is that "management does not like surprises." Since management bears ultimate responsibility for data quality, thoroughly inform them of any problems that may exist. (See Chapter 9 on reporting quality assurance irregularities.)

Bench Analysts

Bench analysts are the employees, degreed and technicians, who perform the actual analyses in the laboratory. Their major responsibility with regard to quality assurance is to comply with the procedures established under the program, and to report to their supervisor any problems that could be related to the quality of their data. Also, encourage them to offer any suggestions they may have for improving the program or solving any problems that might arise.

Support Personnel

Support personnel are those persons, other than analysts, who support the efforts of analysts, supervisors, and managers in the work of the laboratory. They include secretaries, office workers, computer specialists, instrument technicians, and others. Mention has been made and will be made in future chapters of this book of areas where support personnel may be included in the quality assurance program. To the extent that their duties involve quality assurance, their responsibilities are to support the procedures established as part of the program. Examples are the person who maintains the notebook distribution file, or the person who maintains the quality assurance files, both of whom may well be secretaries.

MOTIVATING PERSONNEL FOR QUALITY ASSURANCE

Motivating personnel to follow a good quality assurance program is not difficult if a few simple principles are followed:

1. Communicate to all personnel management's commitment to the program.
2. Be sure that the quality assurance officer has a position of high visibility in the organization, and communicate the fact that he or she has management's backing and authority.

3. Educate employees regarding the objectives of the program, and prepare a written description of the program, so there is little doubt regarding the actual content of the program.

4. Probably most important is to involve all employees in the definition and establishment of the program. Request their suggestions and comments on all proposed procedures. The goal should be to achieve consensus on the procedures to be followed to achieve the desired objectives.

CHAPTER 9

Establishing a Quality Assurance Program

Once management has made the commitment to quality assurance and a quality assurance officer has been appointed it is time to get the program up and running. This chapter will deal with some tried and proven techniques and guidelines to accomplish this. It will be assumed that the laboratory has little or no quality assurance program to begin with and wishes to establish a rather rigorous program.

DEFINING THE QUALITY ASSURANCE PROGRAM

Once the quality assurance officer has been appointed, the next step is to define the details of the program to be established. The quality assurance officer, the laboratory manager, and the supervisors should meet to determine the need for quality assurance and the extent of the work required to achieve it. This will depend on the uses to which the laboratory's data are put, and the principle objectives of the quality assurance program, as described in Chapter 1. Ask questions such as the following and answer them as best as is possible.

What are the consequences of poor-quality data? Are there any undesirable situations that have arisen in the past which could have been avoided by a good quality assurance program?

How often does the lab repeat analyses because clients refuse to accept results when first presented?

Will the laboratory be applying for accreditation by an outside agency, either now or in the foreseeable future?

Are there important company programs, either now or contemplated, that demand quality laboratory data to avoid disaster?

Is the laboratory losing clients, either to other laboratories within the organization or to outside commercial laboratories, because of poor data or a perception of poor-quality data?

What quality control procedures are currently being followed? Are instruments being calibrated frequently, standards being run, control charts prepared, and so on?

What quality assurance procedures are currently in place? Is sample handling satisfactory? Are methods well-defined and authorized? Can results be traced if necessary?

How is the quality of the data generated perceived by the laboratory's clients? If all the lab's clients are in-house, a survey might be taken to determine the laboratory's reputation among clients.

Should the quality assurance program cover all departments of the laboratory, or should the program be customized to suit the different departments?

When these and similar questions have been asked and answered, the quality assurance officer should be able to put together an outline of a quality assurance program based on the laboratory's needs, and those areas in which the laboratory and parent organization is vulnerable to criticism or legal action. Perform an informal audit of the laboratory to ascertain the current quality assurance and quality control procedures.

At this point hold a meeting, chaired by a representative of senior management, with the quality assurance officer and key technical personnel, at which the quality assurance officer can present his or her outline of the proposed program. At this meeting, stress the objectives of the program, and the fact that the cooperation of all personnel will be sought in setting up the program. Consensus at all levels of personnel is the goal. The senior manager, or his or her representative, should stress management's commitment and approval of the outline. The quality assurance officer can then proceed to prepare the written form of the program.

THE WRITTEN PROGRAM

It is absolutely essential that the program be in written form. Simply giving oral instructions to employees will not lead to an effective program. Oral instructions are easily forgotten, mistaken, or may be the source of frequent conflicts because of differing misconceptions of what is intended. These problems may all be avoided with a well-written document. In addition writing a program forces constructive, in-depth thinking about the content of the program.

STANDARD OPERATING PROCEDURES

One proven technique for structuring the written program is by incorporating the instructions for compliance with the program in a set of "Standard Operating Procedures," usually abbreviated "SOPs."

The SOP is not unique to quality assurance. Many companies and other organizations have found that they can simplify their repetitive or standardized operations by means of an operations manual, which is simply a set of SOPs covering a wide variety of topics, for example, how to fill out an expense account, how to calculate vacation time, or what to do in case of a severe snowstorm or other common emergency.

The SOP is a simple, no-nonsense, how-to-do-it document, which leads the reader step-by-step through the operations required to achieve a desired goal. It is not a "policy statement." For example, an organization may have a policy that employees are not allowed to accept gifts from clients. That is a policy statement. An SOP would be a document that tells the employee what to do if a client offers a gift.

Note that in some of the recent literature on quality assurance, especially the EPA protocols, the term "standard operating procedure" is being used to denote the analytical methods. Although the analytical methods may logically be considered a subset of SOPs we will reserve the term analytical method for those testing methods used in the laboratory, to avoid confusion with the type of operating methods being considered here.

Format for Standard Operating Procedures

A standard format for writing SOPs is recommended. It will be of great assistance to the writer and will ensure that nothing is overlooked in the writing. The following is a format for writing standard operating procedures which has been successfully used in several laboratories.

Number. Give every SOP a number so that it can be unambiguously referred to in other documents. If an organization uses SOPs for a multitude of purposes, an alphanumeric designation may be used, such as QA-1 for the first quality assurance SOP.

Title. Use a brief but descriptive title. Remember that the title is the means by which a searcher may look for a given SOP in an index. Therefore, give enough information so that he or she can find the relevant SOP.

Background Information. This section is optional, but include it if the author feels that the reader needs to be informed of the situation that requires the SOP, or which of the quality assurance objectives is addressed by the SOP.

Scope. Delineates the field of application of the SOP to avoid ambiguity over where the SOP is applied. SOPs may be written for limited application in certain situations, and not required for others. An example would be an SOP for handling forensic samples, which would not apply to nonforensic samples.

Purpose. SOPs can be written to cover virtually every task undertaken in a laboratory or any other organization, but most of them would be totally unnecessary. If a good reason or purpose for writing the SOP cannot be found, it is better not to write one.

Definitions. This is also an optional section. Use it to define any terms with which the reader may not be familiar or for defining abbreviations or acronyms.

Operations. This is the core of the SOP. It consists of a series of numbered paragraphs that take the reader step-by-step through the actions necessary to achieve the purpose of the SOP. Take care that the paragraphs are written as clearly and unambiguously as possible, to avoid possible misinterpretation. Describe every step of the operation. Do not assume that the reader knows how to do something just because the writer does.

Avoid words such as "should," "may," or "could" and replace with "shall" or "will." Should, may, and could leave things to the reader's discretion and this may defeat the purpose of the SOP. The SOP is a series of instructions that must be followed, not a set of guidelines that the reader may follow or not as he or she chooses.

SOPs may be long and involved or relatively short. Check long, involved SOPs to see if they can be broken down into several short ones. On the following page is an example of a short SOP covering one topic of a quality assurance program.

Suggested Topics for SOPs

Although each laboratory must, of necessity, write its own SOPs to custom-design the program, the following is a list of some topics that might be included in the list of SOPs. Other topics may be included as necessary.

Laboratory notebooks. How issued, to whom, where stored when completed, how kept (ink, dated, signed, and so on).

Analytical methods. Maintenance of the analytical methods manual, format for written methods, system for numbering methods.

QUALITY ASSURANCE OPERATIONS MANUAL

Standard Operating Procedure QA-4

Title: Reagents and Reagent Solutions

Scope: This operating procedure will cover preparation and storage of reagents and reagent solutions in the laboratory.

Purpose: This standard operating procedure is to ensure good quality control in laboratory operations, and to permit traceability of possible causes of error in analytical results.

Definitions: A *reagent* is any chemical used in a chemical analysis or microbiological test, other than the sample being analyzed. A *reagent solution* is a solution or mixture of chemicals prepared in the laboratory for use in a chemical analysis or microbiological test.

Operations:

1. All reagent chemicals received in the laboratory will be labeled with the date of receipt by the Purchasing Department. When removed from the storeroom, the container will be labeled by the analyst with the date the container was opened and an expiration date, if applicable. The only exception to this rule will be materials used in high volume, for example, extraction solvents, which are used up in a relatively short period of time.

   The dates will be recorded on a special label placed on the container and the label will be placed so that it does not obscure the manufacturer's label. Suitable labels are commercially available in tape form. The common practice of scratching the dates on the manufacturer's label is not good scientific practice and is not acceptable.
2. All reagent solutions when prepared will be labeled with the date of preparation, concentration of active ingredients, and an expiration date. If the shelf life is known or specified in an analytical method the expiration date will correspond to the shelf life. If the shelf life is not known, an expiration data of 6 months from the date of preparation will be assumed. The only exception to this rule will be reagent solutions prepared fresh for each analysis or batch of analyses, and discarded when analyses are completed.
3. All data pertintent to reagent solution preparation and standardization will be logged in a bound notebook kept in the laboratory. Each entry in the notebook will be dated and signed by the person who did the preparation or standardization.

Authorizing analytical methods. Validation of analytical methods, system for authorizing methods, necessary documentation for validation and authorization.

Sampling plans. Need for a sampling plan, factors to consider, and so on. (Separate SOPs may be needed for different matrices.)

Sampling methods. Maintenance of sampling methods manual, placement of methods at sampling point, and so on.

Equipment calibration. Equipment calibration schedules, calibration methods, documentation of calibration.

Equipment maintenance. Necessary preventive maintenance, documentation of maintenance.

Reagents and reagent solutions. Selection of reagents; labeling of reagents with date received, opened, expiration date; documentation of solution preparation, and so on.

Archiving documents and data. What documents should be archived, mechanism for deposit and removal of documents.

Sample handling. Logging, storage, preservation, and identification of samples.

Check standards and control charts. How often check samples are run, how to prepare control charts, how to interpret control charts.

Duplicate samples and control charts. How often duplicates are run, how to prepare *R*-charts, how to interpret *R*-charts.

Reporting quality assurance deficiencies and irregularities. Mechanism whereby quality assurance officer reports Q.A. problems to management.

Running proficiency samples. Procedure for running proficiency samples and reporting results.

Procedure for handling technical complaints. See page 139 for a suggested procedure.

Personnel training records. Documentation of personnel training, where records are kept, and by whom.

Quality assurance files. File system used, where files are kept, who has access to files.

CONSOLIDATING THE PROGRAM

As the quality assurance SOPs are written, circulate them in draft form to the various laboratory supervisors, and other key technical personnel, for their evaluation and input. When the entire program has been codified, call a second meeting for an in-depth discussion of the program.

At this meeting the point should be made that what is not wanted is a "paper program," that is, one which exists only in the written SOPs but does not correspond to the reality of actual laboratory operations. What is necessary is agreement that the SOPs are workable, and that laboratory personnel will put them into practice.

In general two types of objections will be raised. The first type will be that one or more of the SOPs are simply not practical for one reason or another. In this case the quality assurance officer should investigate whether the complaint is justified and, if so, eliminate the SOP or modify it until the objection is removed. If this cannot be done and the quality assurance officer is convinced the SOP is necessary, he or she should insist on changing operations in the laboratory to conform to the SOP or try to achieve the same result by a different means.

The second type of complaint is essentially trivial, based on the common feeling that "We have never had to do it this way before, why should we have to change?" Although the quality assurance officer may recognize the triviality of the complaint it is never wise to dismiss the complaint out of hand. A good quality assurance program will change the way people do things, and it is human nature to resist any change. The quality assurance officer should explain the basic reason for the change in terms of the objectives of the program, and the vulnerability of the laboratory if the change is not made.

It may be tempting for the quality assurance officer to exercise his or her authority at one point or another in these negotiations, and to simply state that this is the way things are going to be done. This is not wise if it can be avoided. The point of this whole exercise is to obtain agreement on the part of the technical personnel so they will put into practice the SOPs as they are written. If the program is imposed on them from above, there will be a temptation to ignore parts of it, or to attempt to foil the system to prove their point that it is unworkable. It may well be necessary to hold several meetings with interim revisions of the SOPs before total consensus on the program and agreement to abide by it are achieved.

MONITORING AND EVALUATING THE PROGRAM

Once consensus has been achieved, set a date for implementing the program. This is conveniently a month or two after the program is in final form, allowing time for educating and training personnel, purchasing any needed material such as notebooks, or setting aside areas for archives and sample storage. The quality assurance officer monitors the efforts at complying to the SOPs and when he or she is convinced that most of the program is in place, conducts an audit of the laboratory.

This audit will disclose areas of noncompliance and may require further revision of the SOPs. If an honest effort has been made to comply, and it is found that compliance is difficult or perhaps impossible, it may be necessary to compromise on the requirements of an SOP.

The next chapter will discuss the various types of audits and techniques for carrying them out. Report the results of all audits to management.

The quality assurance officer should ensure that he or she becomes the focus of the quality assurance program, by monitoring all quality control and quality assurance data generated in the laboratory. Send copies of all control charts to the quality assurance officer, as well as data on proficiency samples, standard analyses, and so on. It is important that laboratory supervisors realize this, and that the quality assurance officer is there to help interpret results and resolve quality problems.

Reporting Quality Assurance Problems

Report all problems, deficiencies, or irregularities in the quality assurance program to management. Although management may not need to take direct action to correct problems, as the person(s) with ultimate responsibility for the data quality they should be made aware of any problems that exist. To ensure that this happens, a formal program for reporting irregularities should be in place.

The following is an example of a quality assurance irregularity report form.

No. ________

Quality Assurance Irregularity Report

Part I (to be filled out by initiator)

Date:

Sample numbers involved:

Nature of QA irregularity:

Signed ____________________

Part II

Steps taken to investigate irregularity:

Probable cause of irregularity:

Steps taken to prevent future occurrence:

Signed ____________________ Date __________

Although usually originating with the quality assurance officer it can theoretically be initiated by anyone in the laboratory. The first part of the form describes the irregularity and is filled out by the initiator. Photocopies of analytical reports, control charts, or other data may be attached to illustrate the problem. The supervisor of the department is required to fill out the second part of the form, which describes the exact nature of the problem, the cause or causes, and steps taken to correct it. The form is then submitted to the quality assurance officer who circulates it to top management for their scrutiny and initialing. It is also the responsibility of the quality assurance officer to follow up to see that the problem has indeed been corrected by the steps taken. He or she should also report back to management on the results.

This system not only ensures that all responsible persons are aware of any problem, but documents that the problem was recognized and corrected, and all responsible persons were notified, which can be of great value in discussing disputed results and determining whether a probable cause of error existed.

Handling Technical Complaints

All analytical chemists know that one of the most disruptive occurrences in the laboratory can be the client who does not believe the results generated by the lab. Many laboratories take the easy way out and simply rerun the sample. If the second answer corresponds to the client's expectations he or she feels justified in making the complaint and is more likely to complain again the next time. On the other hand, if the result confirms the first one, the client often is still not satisfied and the laboratory's reputation may suffer. Many clients who are not technically trained, and some who are, do not understand the concepts of random error in results and the confidence range of a measurement.

It is a wise precaution to have a formal procedure for handling technical complaints. This should be in written form and incorporated in an SOP. Frequent clients can be informed of the procedure and given a copy of the SOP. The following is one way to handle technical complaints:

1. The complaint is handled by a laboratory supervisor, the quality assurance officer, or a trusted, experienced analyst. The first step is to examine the raw data in the analyst's notebook and recalculate the results, to determine if a calculation error occurred, or if the wrong numbers were used in the calculation. Do not assume

that no error could occur because a computer did the calculation. Computers are only accurate when they are fed the correct numbers in the correct program.

2. If no calculation error is found, examine the pertinent quality control data. Was the system in control, as revealed by the control chart? Had there been any maintenance work done on the instrument either before or after the analysis? Had the instrument been calibrated before or after the analysis? Had any quality assurance irregularities been reported at the relevant time?

   In other words, make an honest effort to determine if there was a legitimate cause for an erroneous result. If one is found notify the client of the details, and rerun the sample at no charge to the client, if this is possible.

3. The client may need to be educated to the fact that no measurement is completely error-free and introduced to the concept of the confidence range. It may be that the confidence range of the disputed result includes the value the client thought he or she should have gotten.

4. If no error has been found and the examiner is confident of the validity of the result, and the expected result is outside the 95 percent confidence range of the actual result, an offer can be made to rerun the sample with the following proviso. If the second result is within the 95 percent range of the first, the client will pay for the second analysis. If not, the laboratory will absorb the cost.

WRITING THE QUALITY ASSURANCE MANUAL

Any laboratory with a good quality assurance program must have a quality assurance manual. Laboratories seeking government contracts, or complying with federal or state regulations, or seeking accreditation by an accrediting agency will be required to have such a manual.

In general, a quality assurance manual will contain more information than simply a copy of the quality assurance SOPs. Information on the structure of the laboratory, physical facilities, instrumentation, personnel qualification, and so on will be sought. The manual should be written by the quality assurance officer with the help of management and the supervisors.

Write the manual with the aid of a computer in word-processing mode and store it on a disk, so that revisions and corrections are easily made. Use a loose-leaf binder or similar type of book, again for ease of correction. After completion periodically revise the manual and update it as new material needs to be included. This is done by the quality assurance officer at least annually or whenever major changes have been made in the program.

Many organizations have published guidelines and outlines for writing quality assurance manuals. The following are adopted from the recommendations of Task Force D of the International Laboratory Accreditation Conference, charged with developing guidelines for quality assurance manuals for testing laboratories.

Suggested Outline

1. Introduction
    1.1 *Title page.* Include name of laboratory, name of parent organization, date, revision number, and authorization signatures of senior management.
    1.2 *Table of contents.*

2. Quality assurance policy
    2.1 *Objectives.*
    2.2 *Resources.*
    2.3 *Quality assurance management.* Identify Q.A. officer and staff.

3. The quality assurance manual
    3.1 *Definition of terms.*
    3.2 *Scope.*
    3.3 *Fields of testing covered by manual.*
    3.4 *Management of the Q.A. manual.* Responsibility for manual, procedure for revision, access to manual.

4. Description of laboratory
    4.1 *Identification.* Name, address, type of laboratory, affiliations.

4.2 *Fields of activity.* Types of services offered, samples analyzed, and so on.
4.3 *Organizational structure.* Lines of authority, allocation of functions, and so on.
4.4 *Responsibility for Q.A. program.* Identify Q.A. officer and staff, relationship to other staff.
4.5 *Technical management responsibility.*
4.6 *Quality assurance documents.* SOPs.
4.7 *Deputy assignments.* Persons who assume responsibility in the absence of technical managers and Q.A. officer.
4.8 *Confidentiality of information.* Statement of policy on confidentiality of information regarding clients, analytical results, and so on.

5. Staff
   5.1 *Job descriptions.* Senior and key technical personnel.
   5.2 *Personnel records.* Technical and training records.
   5.3 *Supervision.* Ratio of supervisory to nonsupervisory personnel in each department.
   5.4 *Other relevant documents.* Recruitment policies, training courses, incentive programs, and so on.

6. Testing and measuring equipment
   6.1 *Inventory.* List of major items of test and auxiliary equipment.
   6.2 *Maintenance.* Preventive maintenance (SOPs), maintenance contracts, documentation of maintenance.
   6.3 *Calibration.* Schedules of calibration, calibration procedures (SOPs), documentation of calibration.

7. Environment
   7.1 *Physical facilities.* Floor plans, locations of sensitive instruments.
   7.2 *Heating, ventilation, A/C systems.*
   7.3 *Special environmental facilities.* Sample storage, sample handling, hazardous waste storage, sample disposal, archives, and so on.

8. Test methods and procedures
   8.1 *List of standard analytical procedures.*
   8.2 *Methods for validating nonstandard procedures.*

8.3 *Method of maintaining historical file of methods.*
8.4 *Procedure for authorizing analytical methods.*

9. Updating and control of quality assurance documents
Describe system for periodic updating of Q.A. documents and procedures, and the Q.A. manual.

10. Sample and test item handling
10.1 *System for sample receipt, log-in, and storage.*
10.2 *Sample security.* Precautions against sample mix-up.
10.3 *Sample preservation procedures.*

11. Verification of results
11.1 *System for verifying calculations and data transfer.*
11.2 *Software quality assurance for computerized data software.*

12. Test reports
12.1 *Report format.* Describe system for reporting results, including typical analytical report.
12.2 *Report revision.* Describe system for handling corrections or additions to reports already issued.

13. Diagnostic and corrective actions
13.1 *Feedback.* Describe system(s) for monitoring quality and procedure for applying corrective actions.
13.2 *Proficiency testing.* Describe proficiency testing programs in which the laboratory participates.
13.3 *Interlaboratory comparison programs.* Describe programs for interlaboratory comparison in which the laboratory participates.
13.4 *Use of reference materials.*
13.5 *Technical complaints.* Describe procedure for handling technical complaints.
13.6 *Quality system audits.* Describe frequency of system audits, and procedure for reporting results.

14. Records
14.1 *Record maintenance.* Describe system for archiving raw data, Q.A. records, test reports, calibration and maintenance records.

14.2 *Confidentiality and security.* Describe management policy on confidentiality of records and system for communicating same to employees.

14.3 *Historical file of records.* System for maintaining a historical file of obsolete methods.

15. Subcontracting

Describe system for evaluating quality program of laboratory to which work is subcontracted and policy of reporting subcontracted work to clients.

CHAPTER

10

Quality Assurance Auditing

Once the quality assurance program is in place and up and running it is necessary to periodically review and examine the system to be sure that the program is continually being implemented. Such reviews and examinations are called *audits*, and there are several different types. These are described below.

System audits. System audits are those carried out to compare the operations actually taking place in the laboratory with the written plan for the quality assurance program.

Performance audits. Performance audits are not concerned with the operations in the laboratory, but only with the results of those operations. An example would be analysis of performance evaluation samples, especially if they are blind samples.

Method audits. Method audits are carried out to determine if the laboratory is following the analytical methods as written.

Testing technology audits. Testing technology audits are those carried out to evaluate a laboratory's competence in one or more testing technologies, for example, gas chromatography, GC/MS, or infrared spectrophotometry.

Another way of categorizing audits is on the basis of the purpose of the audit and the persons doing the auditing. In a broad way we can identify "internal" audits, or those performed by the laboratory itself or its parent organization, and "external" audits or those carried out by an external organization, for example, a government agency, professional organization such as the American Industrial Hygiene Association, or an accrediting organization such as the American Association for Laboratory Accreditation.

INTERNAL AUDITS

Internal audits are the means whereby the laboratory reviews and examines itself to verify that the system is following the quality assurance program, or that the laboratory is performing satisfactorily. Internal audits are usually carried out by the quality assurance officer or his or her staff, and results are reported to senior management and laboratory supervision.

Internal System Audits

Internal system audits may be categorized as total or partial with regard to both the quality assurance program and the laboratory. In other words the following combinations may exist:

Total System/Total Laboratory Audits. A total system/total laboratory audit consists in auditing all departments of the laboratory for all provisions of the quality assurance program at one time. If the laboratory is fairly large (20–30 analysts or more) this may take two or more days to accomplish and may cause considerable disruption of the laboratory's normal work schedule. In this case, a partial audit may be more desirable.

Partial System Audit. In this type of audit, only one or two portions of the quality assurance program may be the subject of the audit, with the audit conducted on the whole laboratory.

Partial Laboratory Audit. In this case only part of the laboratory is audited, such as one or two departments, and the audit covers the entire quality assurance program.

Other combinations are obviously possible, such as a partial laboratory/partial system audit. Whichever technique is used, audit the entire laboratory for the entire quality assurance program on at least an annual basis, and more often if the laboratory's needs require it. These audits are conducted by the quality assurance officer and his or her staff, if available.

An easy way to carry out system audits is by means of a checklist. If the SOP system is used to define the quality assurance program, a checklist can be developed for each SOP in the program and used to conduct and record the results of the audit. One copy of the checklist can be used for each department of the laboratory. Report the results of the audit to senior management and the relevant laboratory supervisors. This can be by copies of the checklist, or preferably by a summary report. Management and supervision should sign these reports after reading, and return them to the quality assurance officer for filing. Any discrepancies found between the quality assurance program and actual laboratory practice are corrected by supervisors, or may require a rewrite of one or more SOPs. Corrective action taken is documented and then filed by the quality assurance officer.

The question sometimes arises as to whether this type of audit should be carried out unannounced or whether the laboratory (or at least the supervisor) should be informed ahead of time that an audit will be carried out. Since the audit should not be regarded as a punitive expedition, and since the audit may disrupt normal laboratory operations, never carry it out unannounced. On the other hand, special preparations should not be made by the laboratory staff. The objective is to find out exactly what the laboratory staff is doing, and how well it corresponds to the quality assurance program.

After the second or third audit, the laboratory staff should be in compliance with requirements, and it may be tempting to put off further auditing. This should be resisted. It is important to document the periodic auditing of the laboratory, even if nothing untoward is found.

Internal Performance Audits

These are audits carried out to determine the performance of the laboratory. Obviously all check standards, duplicates, performance evaluation samples, blind samples, and blind duplicates may be considered internal performance audits. In some laboratories NITS standard reference materials are also run on a periodic basis, and these may be considered internal performance audits.

Method Audits

In a method audit, the auditor obtains a copy of the analytical method and observes the analyst performing the analysis. Any discrepancies between the actual performance and the written method are noted. The auditor may be the quality assurance officer or the laboratory supervisor.

Resolve any discrepancies, either by rewriting the method or by instructing the analyst in the proper procedure. If the analyst believes that his or her modification is an improvement on the method as written and there is reason to believe that he or she is correct, then run a validation study of the modified method to ensure that it is at least as accurate and precise as the written method.

Of course, in some cases, as in government-mandated programs, or methods authorized by the client, it may not be possible to use a modified method, and in these cases it is doubly important to use the method as specified. The same is also true of any analyses on forensic samples, especially if litigation is possible.

The frequency with which methods are audited is a matter of professional judgment of the quality assurance officer or the laboratory supervisor. If the laboratory runs a large number of different methods, auditing all of them can represent a large amount of work. On the other hand any samples of special importance, such as forensic samples or mandated compliance samples, are audited at least once at the time of analysis. A good rule of thumb is to audit each method at least once each year.

EXTERNAL AUDITS

External audits are those carried out by an organization other than the laboratory or its parent company. In general, the purpose of the audit is to qualify the laboratory for accreditation by the auditing agency. Federal, state, and municipal agencies may accredit laboratories for compliance to government regulations or for awarding of contracts for analytical work. Professional organizations, such as the American Industrial Hygiene Association, the College of American Pathologists, and the American Society of Crime Laboratory Directors, also accredit laboratories. Large companies sometimes audit, and may accredit, laboratories of their suppliers, and commercial, independent laboratories are often accredited by their clients.

External System Audits

External system audits are generally carried out by the accrediting organization as part of an on-site inspection of the laboratory's qualifications. The quality assurance program is probably the most important part of the inspection, assuming that the physical facilities, personnel, and equipment meet the organization's criteria. A typical assessor's checklist that might be used by an accrediting assessment team for a total system audit is shown at the end of the chapter. This checklist will indicate those areas of interest to an accrediting organization and will give the reader some idea of what to expect in an external audit. This checklist was designed to be applicable to all types of testing laboratories, and some of the items listed may not be strictly applicable to analytical chemistry laboratories.

External Testing Technology Audits

Some accrediting agencies accredit laboratories on the basis of the laboratory's ability to obtain acceptable results within a well-defined testing technology. A testing technology may be defined as a body of analytical techniques centering on the use of a particular instrument (although "wet chemistry" may be considered a testing technology). The accreditation process is designed to evaluate the laboratory's equipment, procedures (including quality control procedures), and analyst expertise to establish the ability to produce quality data. Detailed checklists are usually used to conduct the audit.

External Method Audits

Some accrediting agencies also accredit laboratories on the basis of their ability to perform certain specific tests satisfactorily. An example is the National Voluntary Laboratory Accreditation Program (NVLAP) and the environmental chemistry accreditation program of the American Association for Laboratory Accreditation (A2LA). NVLAP is designed solely for specific test accreditation. A2LA, on the other hand, normally accredits laboratories on the basis of testing technologies, but in the environmental area has been forced, by circumstances, to accredit according to the specific tests mandated by the EPA.

Both organizations carry out a system audit in addition to the specific test audits. Detailed checklists are used in both types of audits.

External Performance Audits

Most accrediting organizations require that the laboratory perform satisfactorily on performance evaluation samples, which are periodically submitted to the laboratory for analysis. Initial accreditation and continued accreditation usually require acceptable performance on these samples. It is wise to be aware beforehand of the criteria on which acceptability is determined.

A common scheme is to require that the laboratory's results fall within the 95 percent confidence range established by other laboratories in the program or by a reference laboratory. If the result falls outside the range, the laboratory is asked to analyze another sample, since this can occur by random variation 5 percent of the time. If the second result falls outside the limits, accreditation is usually denied, since the probability of this happening by chance is $\frac{1}{400}$ ($\frac{1}{20} \times \frac{1}{20}$). Accreditation is then withheld until the laboratory can demonstrate acceptable results.

Another form of external performance audit of which the laboratory may not even be aware is blind samples submitted by clients. Laboratory clients are becoming more knowledgeable about quality assurance, and some are submitting samples of known composition as a check of the laboratory's ability to do quality work. Another type of check may be blind duplicates, prepared by simply splitting a sample in half and submitting each half as a separate sample.

Some laboratory managers and supervisors become quite defensive when they learn that clients are sending in blind samples or duplicates. This is not an appropriate attitude. It is far better to encourage clients to do this and then share the results with laboratory management or supervision. There is no better way for the laboratory to determine the true run-of-the-mill accuracy and precision of data than this type of blind-sample program.

NOTE

Typical Assessor's Checklist[1]

1. Name and address of laboratory
Technical manager
Quality assurance officer
(The following should be answered with yes, no, or n/a.)
Satisfactory organization structure
Can perform representative tests
No undue pressure observed
Staff responsibilities are clear
Adequate security for client data

2. Quality System
2.1 Laboratory has internal Q.A. system
Q.A. manual available
Manual maintained regularly
2.2 Q.A. manual contents
Organization charts
Staff duties pertaining to quality
General Q.A. procedures
Q.A. procedures specific to each test
Proficiency/reference materials used
Feedback and corrective action program
Complaint handling procedure
2.3 Q.A. system periodic reviews recorded

1. For total system audit (adapted from *ISO Guide 25-1982*).

3. Staff
- 3.1 Necessary education, training, and knowledge
- 3.2 Job descriptions available for key technical personnel
- 3.3 Adequate supervision
- 3.4 Technical/Q.A. staff deputies (i.e., replacement personnel for staff when not present)
- 3.5 Qualifications, training, and experience recorded

4. Testing and measuring equipment
- 4.1 Equipment available for scope of tests
- 4.2 Equipment maintained and instructions available
- 4.3 Overload and mishandling procedures available
- 4.4 Equipment records maintained:
  - 4.4.1 Name of equipment item
  - 4.4.2 Manufacturer's name, type, serial number
  - 4.4.3 Dates received, placed in service
  - 4.4.4 Current location
  - 4.4.5 Details of maintenance
- 4.5 For measuring equipment:
  - 4.5.1 Date of last calibration and reports
  - 4.5.2 Maximum time between calibrations
- 4.6 Calibration labels used?

5. Calibration
- 5.1 Initial calibration and calibration program
- 5.2 Traceable measurements (where applicable)
- 5.3 Reference standards for calibration
- 5.4 Reference standards calibrated appropriately
- 5.5 In-service testing equipment checks
- 5.6 Reference materials traceable
- 5.7 Written calibration procedures available?

6. Test methods and procedures
- 6.1 Equipment operating instructions
- 6.2 Methods are as required by specifications
- 6.3 Nonstandard methods fully documented
- 6.4 Calculations and data transfers checked
- 6.5 Data processing accuracy checked
- 6.6 Historical file of obsolete methods kept?

7. Environment
- 7.1 Equipment protected and environment monitored

7.2 Access to test areas controlled (as needed)
7.3 Adequate housekeeping

8. Handling of samples
8.1 Sample identification procedures adequate
8.2 Bonded storage available (if needed)
8.3 Sample protection procedures adequate
8.4 Rules for receipt, retention, and disposal of samples

9. Records
9.1 Records adequate to permit repeat of test?
9.2 Records and reports secure?

10. Test reports
10.1 Work in laboratory covered by test reports
10.2 Each test report contains:
Name and address of laboratory
Sample identification and description
Unique identification
Name and address of client
Date of receipt of sample and date tested
Test results related to sample
Identity of method used
Description of sampling procedure
Any deviations, additions, and so on to test
Identity of any nonstandard test used
Test results
Measurement uncertainty (if relevant)
Signature and date
Statement regarding reproduction of report
Subcontracted work identified?
10.3 Report format OK?
10.4 Supplemental procedures

CHAPTER 11

Laboratory Facilities and Quality Assurance

Certainly one of the most important factors in the analytical system impacting on the quality of analytical results is the physical facilities available in the laboratory. It is well recognized today that controlled temperature, good lighting, adequate electrical supply, and a pleasant working environment are necessary in today's laboratories. Two general principles are involved: the proper environment for optimum operation of equipment and instruments, and the proper environment for optimum performance of the employees.

This chapter discusses some of the requirements of laboratory physical facilities that will enhance the quality assurance program of the laboratory.

GENERAL REQUIREMENTS

Some of the physical facilities necessary for proper operation of today's sophisticated instrumentation were discussed in Chapter 6. It was pointed out that an adequate heating, ventilation, and air-conditioning (HVAC) system was mandatory. To recapitulate some of the needs for instruments:

1. If possible, a separate room for housing the more delicate instrumentation and equipment is desirable. This will minimize contamination from fumes and dust.
2. In choosing an instrument room, take care to avoid excessive vibration. In other words do not consider a room fronting on a busy highway with heavy truck traffic.
3. If a separate instrument room can be established, give consideration to installing a separate HVAC system for this room. This will allow a slightly higher air pressure to be maintained in the room and help to keep out fumes and dust. Keeping the door closed is mandatory.
4. Electric service should be adequate to the room, and if possible, install voltage surge protection, otherwise separate surge protectors will need to be provided for each instrument.
5. Adequate electrical grounding will be needed for good instrument operation.
6. Depending on the laboratory's field of operation, consideration might be given to obtaining an emergency power source, for example, a generator, for power blackouts. Apply this to the instrument room, computer room, refrigerated sample storage areas, incubators (for microbiology), appropriate instruments, and other areas where power interruption may cause loss of data, or the need to rerun analyses. Operation of the emergency generator should be automatic when power loss exceeds a preset time interval, for example, 10 seconds.
7. Another option available, especially for computers, is an uninterruptable power supply (UPS). This device is a battery backup system, which permits continued operation for a limited time, for example, 10 minutes or an hour, usually sufficient time to back up any data in the computer before shutdown.
8. Have fire extinguishers of the "Halon" type, which do not leave

a residue when used, in the instrument room. The usual type ABC extinguisher leaves a powder residue that can ruin an instrument.

SAMPLE-HANDLING FACILITIES

Sample Receipt and Log-In

In small laboratories, handling 10 to 20 samples per day, little is needed in the way of sample receipt and log-in, other than a desk or table and a log-in manual or computer terminal/work station. However, in larger laboratories, especially if the sample load approaches 100 per day or more, give some consideration to a separate area for sample receipt and log-in. This is especially true if the laboratory is engaged in trace or ultratrace analysis, or in handling especially hazardous materials.

The accountability objective of quality assurance requires that the samples be handled in such a way as to avoid mix-up and to ensure that the sample when analyzed has undergone minimum chemical and physical change while in the laboratory. This may require a refrigerator in the sample receipt area, possibly a freezer, special light protective containers, inert gas, or other facilities to prevent sample change.

Samples for trace or ultratrace analysis, such as environmental samples, may require special precautions to avoid contamination. One sample with a high concentration of volatile analyte can contaminate a room full of trace analysis samples, especially if the sample container is opened or accidentally spilled in the sample receipt area.

Special precautions must also be taken if extremely hazardous materials are present in the samples. It may be necessary to have a fume hood in the sample receipt area and require employees to use protective devices such as gloves or face masks.

The sample receipt area can also serve as a collection point for samples which require special disposal procedures.

Sample handling can be a rather technical job, with its own needs for equipment and techniques. It is for this reason that it is recommended that sample receipt, log-in, and storage be limited to certain personnel only, who receive the necessary training to do the job properly. Such training should be duly documented in personnel files.

Sample Storage Facilities

Very few laboratories are so equipped and staffed that samples are analyzed immediately upon receipt. In most cases samples must be stored between the time of receipt and the time of analysis. Once again the quality assurance objective of accountability demands that samples be stored in such a manner that there is minimal change in composition or physical state before analysis. It is also necessary to ensure that samples cannot be inadvertently mixed up or lost.

In many laboratories samples are simply transferred from the sample receipt and log-in area to the laboratory, and stored in the laboratory until analyzed. This is not a recommended procedure. Storage in the laboratory exposes the samples to the possibility of contamination and increases the chances of sample mix-up.

A far better system is to set aside a separate area for sample storage prior to analysis, preferably a small room. The room may contain one or more refrigerators for samples requiring refrigerated storage, and, along the walls, a series of shelves for storage of other samples. The shelves should have a series of partitions, creating a nest of cubbyholes for sample storage. Each cubbyhole should have a number or alphanumeric designation, for example, B2, designating the second row in the second column.

Persons handling sample receipt and log-in should indicate in the log the location where the samples are to be stored. A daily printout of this information is given to the laboratory supervisor so that the analysts will know where to find a given sample. If computerized log-in is used this can easily be handled by the computer. The computer can also keep track of which storage locations are empty by the sample numbers of those samples that have been analyzed and reported out. A weekly printout of those samples that should be discarded can prevent the buildup of analyzed samples that should be discarded. If the laboratory handles "forensic" samples, that is, samples that may be important in a legal or other type of dispute, special sample storage facilities may be necessary. The legal concept of chain of custody applies here. Store the samples in a secure, locked, storage area with limited access, prior to analysis. "Limited access" means that only one or two persons have keys to the locked storage area.

Sample Disposal

In today's world of concern over environmental pollution and hazardous waste there is no place for the old sample disposal systems of pouring samples down the sink or dumping them in the trash pickup. Today's laboratory manager must make sample disposal a priority item in his or her list of things to control.

Once again set aside a separate area for sample collection and disposal, and use trained personnel for the task. Safety considerations must be paramount in most laboratories since most chemical samples are hazardous to a greater or lesser degree. Environmentally safe disposal procedures must also be developed.

The difficulties inherent in disposing of hazardous samples have led some laboratories to establish a policy of returning all samples to their clients after analysis. This is a wise policy, if the laboratory's clients accept it. Forensic samples may require special disposal handling. In some cases samples must be stored for long periods pending a trial. In this case, "bonded storage" may need to be considered as an option.

GLASSWARE-CLEANING FACILITIES

Many laboratories have established separate glassware-cleaning facilities. This not only improves efficiency in the laboratory, but can have a positive impact on quality assurance, by minimizing a common source of contamination. If trace analyses are run in the laboratory, set up separate facilities for cleaning glassware used in this work, and special techniques may need to be devised. For example, some laboratories have found that a high-temperature baking of this glassware can significantly reduce blank values in trace organic analyses. Personnel working in this area may need special training, and again, for quality assurance purposes, document this training.

ARCHIVES

The need for archiving raw data, quality control data, quality assurance reports and documents, and other pertinent documentation has been mentioned in previous chapters. In many laboratories this need is not fully recognized—filled notebooks are kept in the analysts' desks, reports and documents in office file cabinets with open access, and so on. The quality assurance objective of traceability, that is, the ability to establish the actual data taken and the condition of the analytical system at the date an analysis was run requires that this material be easily found. This requires that an archive area be established.

The archive area should be clean, dry, and not subject to extremes of temperature. Do not use attics and basements. Attics tend toward extremes of temperature, and basements are often high-humidity environments subjecting paper documents to mold and condensed water. Do not establish the archive in the near vicinity of a laboratory because of the dangers of fumes, fires, and explosions. Make periodic inspections for insect and rodent damage and take corrective action if needed.

Limit access to the archives, that is, only one or two persons should possess the key. The archivist or keeper of the archives maintains an index file of all materials deposited in the archives, and a log of materials removed, by whom, and when returned to the archives. In this way the location of all materials archived can be established at any time.

Depending on the laboratory's needs it might be advisable to microfiche all material in the archives and keep the microfiches at another location, or at least in another building. This will minimize losses in case of fire at the laboratory.

The question of how long to keep archived material can be a troublesome one, especially if archive storage space is limited. There is no hard and fast rule, since this depends on the laboratory's place in the scheme of things. The EPA requires 10 years for data in support of environmental regulations, and the FDA mandates 7 years after a new drug application, which is also about 10 years since several years often pass between the time the work is done, and the time of the application. A good rule of thumb then would be 10 years, or longer if the archive space limitations permit.

CHAPTER

12

Computers and Quality Assurance

In the past decade one of the most important technological innovations in the analytical laboratory has been the widespread introduction of computers. This has followed the development of low-cost computer hardware, especially the ubiquitous personal computer or PC. For a few thousand dollars the laboratory can now purchase computing power which formerly would have cost hundreds of thousands or millions of dollars.

The analytical laboratory as a generator of information, especially numerical data, is, of course, a natural place for the use of computers, since a major function of computers is the collection, storage, calculation, manipulation, and output of data in a variety of forms. It would be difficult today to find a laboratory without at least one computer.

The hardware systems in use today in laboratories run the gamut from one or two PCs through multiple PCs connected to each other through a local area network (LAN), to PCs or dumb terminals connected to a minicomputer or a large mainframe. The particular system in use seems to be a function of the evolution of computer uses in the laboratory, the size of the laboratory, and the availability of computer power from the parent organization of the laboratory.

In the first edition of this book readers were warned against beginning with one or two PCs in the expectation of being able to grow by adding computers as needed. Like so many precepts in the computer field, this one proved to be unfounded, since many laboratories have done just that, and hardware and software suppliers have supported this trend. The development of the LAN concept and the necessary software, peripheral devices, and hardware (e.g., connectors and "boards") made this possible.

However, the LAN, while capable of acceptable performance for data collection, is less than optimum for other computer functions, for example, data storage, manipulation, and graphing. When the size or complexity of the laboratory's work reaches a critical point, the use of a minicomputer as the base of the system will improve efficiency and scope of the applications of the computer. The individual PCs can then serve as terminals for the minicomputer. The use of the parent organization's mainframe computer is not recommended if it can be avoided because then the laboratory is at the mercy of the mainframe, and the data-processing personnel managing it, who are often unsympathetic to the laboratory's needs.

LABORATORY INFORMATION MANAGEMENT SYSTEMS

The first, tentative steps to the use of computers in the analytical laboratory focused on attempts to use them for data acquisition from electronic instruments, and subsequent reduction of the data to summary form. However, the development of inexpensive microprocessors soon made it possible to transfer these data-handling processes to built-in microprocessors offered as part of the instruments themselves.

With the development of low-cost computer power (including PCs) came the realization that computers were useful for much more than just "number crunching"; that they could be used for data storage, graphic displays of data, decision making (given well-defined criteria), organization of data into tables, and word processing. In other words, computers could be used for "information processing."

Thus was born the concept of the laboratory information management system or LIMS.

In computer science terms a LIMS is a software application that operates on a "database." The database consists of the samples logged into the system and the analytical results on those samples. The function of the software is to store information on those samples, calculate data from the inputted data, manipulate data, and output data in the form of reports, graphs, tables, and other summary data. The software/database combination can be used for many other purposes as well, such as preparation of workload lists, client lists with associated worklists, and financial data on laboratory costs.

The subject of LIMS is a very complicated one and clearly beyond the scope of this book. For detailed information on LIMS acquisition, functions, availability of commercial software, and so on, the reader is referred to the excellent book by Mahaffey.[1] Those aspects of LIMS operations that affect laboratory quality assurance are the subjects of the remainder of this chapter.

LIMS Functions and Quality Assurance

The following is a list of common LIMS functions available in most LIM systems and how they impact on quality assurance.

Sample Log-In

Sample log-in is one of the most important functions of a LIMS. The basic principles that should apply to sample log-in, whether computerized or manual, were discussed in Chapter 7. For quality assurance the following should be available in a LIMS:

Access. Only designated personnel are able to log in samples, that is, program the LIMS to grant access to the log-in function only to the personnel with the proper password.

Sample Identification Number. The LIMS generates and assigns sample numbers to samples as they are logged in. An advantage is a LIMS that automatically prints a sample label at the time of log-in, which can immediately be attached to a sample container. Some LIMS generate a bar-code symbol for the label, which incorporates the sample number and which can be read by analysts with a light pen or similar device to ensure they have the proper sample.

Computer-generated sample numbers will prevent the accidental assignment of the same number to two samples, or the skipping of sample numbers in a sequence, both of which can occur in manual sample log-in.

Data Backup. Periodically reduce computer sample log-in data to hard copy, at least daily, and more often if volume of samples warrants it. Never rely on the computer for long-term storage of data, because data in the computer are ephemeral—they can be lost through power outages, "head crashes," or other catastrophes. Remember Murphy's law! (If anything can go wrong it will.)

Hard-copy printout of the sample log-in data will ensure that a relatively permanent record exists of the receipt of the sample.

Calculation of Analytical Results

This is a function that is not available in many commercial LIMS but is highly desirable if it can be had. Most LIMS require that results be entered into the computer after external calculation. If the computer can be programmed to do the calculations, a common source of error can be eliminated.

In order to achieve this, however, it is necessary that the analytical method be identified, and programs written to accept raw data, calculate according to the appropriate equation, and store results. This is one good reason for numbering all analytical methods as recommended in Chapter 5.

Although calculation errors can be eliminated in this way, it does not eliminate errors due to incorrect input of data. A common error is to transpose digits in a number, such as 1243 instead of 1234. One method of minimizing these errors is to require that all data be entered twice, that is, the computer will not calculate until the data are entered the second time and the two entries correspond. Of course, this requires twice the data entry time and work.

Another method of at least identifying this type of error is to store the raw data and later print it out as a record of correct input to the calculation program. In one system the author is familiar with (developed in-house), the raw data are printed on the back of the analytical result form and filed. The front of the form, which contains the calculated results, is photocopied and sent to the client. At any future date, the filed copy can be checked against the analyst's notebook to see if the proper data were entered into the computer. This system also identifies, by employee number, the person who entered the data in the computer.

The "electronic notebook" concept was discussed briefly in Chapter 7. In this type of system a PC is used for direct data acquisition from instruments, on which calculations can be made, and the raw data and calculated results dumped into the LIMS. This has the advantage of direct data acquisition without tying up the LIMS for long periods of time.

Sample Status and Verification

A good LIMS will keep track of a sample's status and not release results until the final result is verified. In general, four status levels are recognized: received and awaiting analysis, analysis in progress, analysis completed but not verified, and analysis completed and verified. A fifth status might be analyzed and reported out, but many LIMS simply report results after verification, and then erase the sample from computer memory to conserve memory.

Verification usually means that a laboratory supervisor or other authorized person has looked at the results (and sometimes the raw data) and found them not unusual. If the results are not approved, the sample may revert to the unanalyzed status, awaiting a retest.

In some laboratories access to the LIMS is given to selected clients to enable them to receive sample results by printout at their home locations, via a telephone modem and computer terminal, without waiting for hard copy from the laboratory. Only allow this type of access to samples in the analyzed and verified status, and program suitable safeguards into the computer to prevent unauthorized access.

The desirability of a sample storage area, where samples may be held while awaiting analysis, was described in Chapter 11. If this facility is available, the LIMS can be programmed to assign sample storage location during the log-in procedure. To do this, a database must contain the available storage locations and their status, that is, whether occupied or not. By keeping track of which samples have been analyzed and reported out, the LIMS can prepare a daily or weekly list of samples to be discarded, and update the sample storage base accordingly. This will also serve to minimize the number of analyzed samples in the laboratory that are to be discarded by reminding the lab supervisor periodically of this chore.

Reporting Results

All LIMS have programs for reporting results. The general principles for this function have been described in Chapter 7 under Analytical Reports. A desirable feature is the system whereby the computer assigns a code number to each report that designates which analytical method was used, and, in some cases, the personnel number or initials of the analyst. This will speed searching for traceability purposes.

Control Charting

Maintaining control charts for analyses is one of the most valuable functions of a LIMS, since manual charting can be a labor-intensive chore, if the number of analytical methods in use exceeds a few. The following describes a system with which the author is familiar, and which works very well.

When the analyst analyzes a check sample for control charting, a special number is assigned to the sample that flags it for the computer as a check sample. The data entered against the sample number are stored in the database for the control chart. The computer continually calculates the mean and standard deviation as each new sample is added to the database.

If the new sample result exceeds the 95 percent confidence limit (taken as two standard deviations), the analyst is informed of the fact and instructed to inform his or her supervisor. Note that, once entered, the data for the sample cannot be changed in any way. This keeps the supervisor informed of any possible problems in the analytical system.

When requested, the computer displays, and prints out if requested, a control chart along with the following information:

- The name and number of the analytical method
- The mean and standard deviation calculated for all replicates run on the check sample to date

- A list of all results obtained since the last printout, the date run, and the personnel number of the analyst
- The mean and standard deviation of all samples since the last printout
- A control chart showing the last 25 samples that have been run, with lines showing the mean and two standard deviation limits
- A message giving the number of samples exceeding the two standard deviation limit since the last printout, and for the year to date

With a system such as this the LIMS manager can obtain a monthly printout for all analyses under the control charting system, with copies for each laboratory supervisor and the quality assurance officer. The quality assurance officer and supervisors can easily scan each chart for anomalies, determine if the accuracy or precision has changed significantly, and identify analysts who may have a bias toward high or low results, or who may be responsible for a significant number of blunders.

Other Functions

A good LIMS can offer many other functions, not necessarily related to quality assurance, but of great value in laboratory management. For example, worklists can be prepared and printed out on a daily basis, listing all samples currently in the laboratory, the status of each sample, and the analyses needed. This can be of great assistance to laboratory supervisors in organizing the workload and predicting future shortages or excesses of manpower.

The worklist can also be used to analyze sample-holding times to give managers an idea of average turnaround time, number of samples exceeding a given turnaround time, or to identify samples that are approaching or have exceeded a planned turnaround time.

In conjunction with a time card or time allotment system the LIMS can be of help in deciding the cost of certain routine analyses and charges to be levied against clients. The client database can identify those clients responsible for the greatest amount of the laboratory's work, or list clients in order of workload. This type of client list analysis can help to plan future manpower requirements for the laboratory.

Managing the LIMS

In the words of Mahaffey:[1] "A LIMS doesn't come with an on button." In other words, it is not enough to simply purchase hardware and software, put the two together, and hope that laboratory employees will use it. Instead the selection, installation, and implementation of a LIMS requires very careful consideration based on goals and resources, and must be followed by education of the laboratory staff in the proper use of the LIMS.

This means that the steps in establishing and maintaining a LIMS must be a management function. Since the LIMS can be expected to be an on-going aspect of laboratory management it is necessary to designate a LIMS manager.

The LIMS manager should be a person with a strong background in analytical chemistry and more than a passing interest in computer programming. He or she must be thoroughly familiar with all of the technical operations of the laboratory, and should work closely with the quality assurance officer in setting up the system.

Selecting a LIMS

Selecting a LIMS for a given laboratory is a difficult task, depending on the laboratory's mission, financial resources, and availability of "computer literate" personnel. There are three recognized routes: in-house development, customized written software, and purchase of available commercial software. The in-house route is not usually taken, because it requires in-house personnel with the requisite computer programming skills, as well as knowledge of analytical laboratory work. Hiring a consultant to write the software can be risky and expensive, and is probably not a viable option, unless the laboratory has unusual requirements. Choosing a commercially available LIMS is probably the safest and best alternative if one can be found that meets the laboratory's needs. A consultant can be invaluable in advising the laboratory in this situation. The Mahaffey book is an excellent review of commercial LIMS available at the time of publication (late 1989).

Two important factors to be considered are the degree of vendor support for the software and the availability of vendor education for laboratory personnel. For evaluating the quality of the software, the book by Beizer[2] is highly recommended. The Beizer book is a good discussion of the factors that distinguish good software from bad (all kinds of software).

LIMS Documentation

A LIMS represents a large software program, usually consisting of hundreds or thousands of lines of computer code. Ideally, it would be desirable to have hard-copy documentation of this software, because a program that large can easily harbor errors or "bugs" that may not become apparent until after long periods of use. If the LIMS has been developed in-house, or by a hired consultant, the hard copy is easy to obtain, but this is not the case for commercially purchased LIMS. This is understandable because the software manufacturers must take steps to guard against unwarranted copying of their product.

However, many commercial LIMS allow the user to incorporate user-written subprograms into the system. This would be necessary if the user wishes to have the computer calculate analytical results, for example, since the algorithm must depend on the method used, units of raw data, and so on. This may also be necessary to some extent for computer-generated control charts. In fact, virtually all the functions listed previously may require some user-written code to customize the LIMS to the particular laboratory's needs.

All such laboratory-generated programs must be documented. Include the date the copy was made, the name and signature of the programmer, and the revision number, if appropriate, in this documentation.

Other necessary documentation includes any auditing or validation runs made on the LIMS. Make these runs periodically and document them as proof that the LIMS was functioning correctly at the time the run was made.

Software Quality Assurance

A LIMS and the laboratory in which it is used represent a complex system for manipulating information by means of the motion of electrons guided by subminiature magnetic domains. As such, the system is a *delicate* one, subject to a variety of environmental interruptions. In other words, we must always be aware of the possibility of deterioration of software quality.

A LIMS is a very complex computer program, and it rarely happens that a LIMS can be installed in a laboratory without any bugs. Some of these bugs are immediately apparent and easily fixed by laboratory personnel, or a representative of the software producer. On the other hand there may well be bugs present that are not readily apparent and that only show up after long use of the program, or under unusual circumstances. This type of problem is of great concern to quality assurance personnel, but there are no hard and fast rules for detecting bugs. The reader is referred to the book by Schulmeyer and McManus[3] for further discussion of this topic. Although the book does not treat LIMS specifically it is a good reference for the subject of software quality assurance.

It is good management policy to establish a validation protocol, or audit protocol, to be applied to the LIMS periodically to demonstrate that the software is operating correctly. Check all important functions of the LIMS, using dummy data, and file and archive results as hard copy. It is important to validate the LIMS under conditions similar to daily operations. In other words do not do the validation on weekends or evenings when the system is not busy. The validation report is addressed to the quality assurance officer by the LIMS manager.

LIMS Security

The laboratory/LIMS system is a sensitive and vital part of laboratory operation, and therefore must be protected from unauthorized access and the possibility of tampering, theft, or destruction of data. Physical protection of the hardware is essentially the same as for any other valuable equipment. The needs for electrical protection and protection of the computer hardware in case of fire were discussed in Chapter 11.

Protecting the data in the LIMS from unauthorized access is a common feature of all commercial LIMS. Usually a hierarchy is defined and access to the various functions of the LIMS depends on the user's place in the hierarchy. The LIMS manager has access to all functions of the system as do some, if not all, senior managers. On the other hand, an employee in the sample receipt and log-in group may only have access to the functions relevant to his or her job. An analyst may have access to the result-calculation function, but not verification of results, while the quality assurance officer may have access to the control chart display function. In other words each employee has access only to those functions relevant to his or her position.

The functions that a given employee may access are generally defined by his or her password, which must be entered into the computer whenever the employee is on-line. This accessibility must be programmed into the LIMS and is given careful consideration by the LIMS manager.

While the password system may be considered an adequate security measure, and is undoubtedly effective with the majority of laboratory employees, it must be realized that it does not take much for a determined person (such as a disgruntled employee) to break the system and cause damage. Educate employees about the importance of their passwords and the necessity to keep them confidential, even from their friends and fellow-employees.

The LIMS manager issues passwords, or possibly the computer itself does (e.g., a random four-digit number). Do not allow employees to choose their own passwords, because invariably they will choose one they can easily remember, such as their initials or their birthdate, neither of which is very confidential information.

Computer Viruses

In the past few years the problem of computer viruses has received a great deal of attention in the media. A computer virus is a small program or segment of a program, surreptitiously introduced into a computer program, that continually replicates itself, and may or may not destroy data on the disk. If undetected it may completely ruin all the data on a disk, and infect other disks where it will continue its destruction.

Fortunately the computer virus problem has not been experienced by the analytical laboratory community to any great extent. Nevertheless, the potential for the problem does exist, and laboratory managers should be aware of the possibility. The book by Fites, Johnston, and Kratz[4] is recommended as a good discussion of the problem, with suggestions on avoiding it and correcting the problem if it occurs.

The best way to avoid the problem is to control the routes by which viruses enter the computer. The most common of these is by introduction of software that has been downloaded from a public "bulletin board system" or BBS, or that has been obtained from a source other than a reputable software producer. These are usually games or utility programs, but could also be programs such as word processors or spreadsheets. The widespread use of PCs as computer terminals may tempt employees to run these programs on their personal terminals. To avoid this virus source, prohibit laboratory employees from using programs other than those authorized by management.

The other possible source is in purchased software that has been contaminated at the manufacturer. While this has happened in the past, it is probably safe to say that software manufacturers are now thoroughly aware of the problem and it is probably not a likely source today. Nevertheless, back up all newly purchased program software, and write-protect the original and backup disk. This will at least permit the localization of the problem should it occur.

LABORATORY ROBOTICS

In addition to LIMS, the last decade has seen computers penetrating laboratory operations in the form of robots, or the application of *robotics,* the science of robot performance. Laboratory automation is not new, in that we have had automated instruments, peripheral devices such as automatic sample injectors, and automatic machines for performing certain well-defined analytical procedures such as spectrophotometry for many years.

However, these devices are all limited in application, and are concerned chiefly with the measurement stage of chemical analysis. The robot, on the other hand, enables the laboratory to automate the sample preparation stage in the analysis. In addition, robots are computer controlled and can be programmed to perform a variety of functions, and to vary their performance based on the decision-making capability of the computer.

The major impetus toward the introduction of robots is undoubtedly the replacement of human labor in chemical analysis, and the financial savings possible. Labor is still the biggest cost item in the analytical laboratory, and anything that will reduce cost without sacrificing quality will be welcomed and will prevail. Automated and computerized instrumentation has reduced the labor cost of the measurement step, but the sample preparation step(s) remain a labor-intensive operation. Although the use of robots in laboratories is still in its infancy, there is little doubt that it will grow rapidly.

In addition to lowering costs there are other advantages to the use of robots. Robots can be used in hazardous environments, and can use hazardous materials, adding to laboratory safety. They enable more efficient use of expensive instrumentation and will work 24 hours a day for indefinite periods (provided they don't break down).

Robots in today's laboratory are essentially electromechanical devices, controlled by a program installed in a microprocessor. The robot can be used to implement most laboratory unit operations used in sample preparation, such as weighing, dilution, solution, mixing, pipeting, grinding, and addition of reagents. Once programmed for a series of such tasks the robot will perform them repetitively as long as there are samples to be prepared. In addition, the series can be varied, using the decision-making function of the microprocessor. For example, if a sample contains a level of analyte higher than can be accurately measured, the robot can divert the sample back to the beginning and weigh out a smaller sample, or dilute a solution to the point where accurate measurement can be made.

There are certain aspects of robotic operations that bear on laboratory quality assurance. Robots are said to operate with greater uniformity than humans, and this leads to increased precision in analyses. While this is theoretically true it would be prudent on the part of the laboratory supervisor to obtain hard data to support this claim. It should also be born in mind that robots are partly mechanical in nature and that mechanical devices are subject to wear and corrosion. Therefore, the fact that the robot delivered the exact volume of liquid when it was first set up in a procedure does not mean that it will continue to do so after several thousand analyses have been run. In other words, the need to monitor the analytical system does not diminish because a robot is part of the system. Still use control charting and check samples, and periodically calibrate those parts of the robot system involved in measurement, such as automatic pipetors, and document the calibration.

Another advantage claimed for robots is uniformity of analysis, that is, the elimination of analyst-to-analyst variation. It is well known that human analysts seem to have built-in biases, which lead to different results even when the same method is used by different analysts. The robot will slavishly follow the programmed method until programmed differently thus ensuring that all samples receive equal treatment. An incidental advantage is that a printout of the program will provide positive documentation of the method used.

A third advantage claimed for robotics is the ability to transfer methods from one laboratory to another without loss of integrity, provided the two laboratories are using the same robotic system and the same program. This feature is a real advantage in method development and in multilaboratory organizations.

Robotics appears to be the final step in the long-awaited automatic laboratory, with samples going in at one end and analytical results coming out the other. However, keep in mind the cautionary tales of our brothers in the humanities, such as the *Sorceror's Apprentice* or the computer/robot HAL in *2001 A Space Odyssey*.[5]

REFERENCES

1. Mahaffey, Richard R. *LIMS Applied Information Technology for the Laboratory*, New York: Van Nostrand Reinhold, 1990.
2. Beizer, B. *The Frozen Keyboard: Living with Bad Software*, Blue Ridge Summit, Pa.: TAB Books, Inc., 1988.
3. Schulmeyer, G. Gordon, and James McManus. *Handbook of Software Quality Assurance*, New York: Van Nostrand Reinhold, 1987.
4. Fites, Philip, Peter Johnston, and Martin Kratz. *The Computer Virus Crisis*, New York: Van Nostrand Reinhold, 1989.
5. Hurst, W. Jeffrey, and James W. Mortimer. *Laboratory Robotics*, New York: VCH Publishers, Inc., 1987.

CHAPTER

13

Mandated Quality Assurance Programs and Laboratory Accreditation

The basic principles of laboratory quality assurance, with detailed descriptions of how to implement them in the modern laboratory, have been described in previous chapters. Such implementation comes with a price, and it is the function of the laboratory manager to decide the extent of the rigor of the program for his or her laboratory, consistent with the laboratory's needs and the financial constraints.

However, more and more managers are finding that, due to circumstances, they do not have an option: To satisfy their clients needs they must establish a rigorous quality assurance program as defined by an outside organization. Federal, state, and even municipal agencies to which laboratories submit data to demonstrate compliance to relevant laws and regulations are requiring that laboratories have, and can demonstrate that their data is generated under, adequate quality assurance programs.

Legal departments in many organizations are demanding that

analytical data that may be used in litigation be generated under a good quality assurance program. For example, data relating to employee exposure to toxic chemicals in the workplace will have greater credence in labor disputes if the laboratory is accredited by the American Industrial Hygiene Association.

Laboratory accreditation has been defined as "verification by a competent, disinterested, third party that a laboratory possesses the capability to produce accurate test data, and that it can be relied upon in its day-to-day operations to maintain high standards of performance."[1] The independent commercial laboratory may find that accreditation in one form or another is almost indispensable to its existence. In addition, many large corporations are requiring that laboratories of their suppliers be accredited.

Some of the more common "mandated" quality assurance programs, and also the idea and implementation of laboratory accreditation, will be discussed in this chapter.

THE GOOD LABORATORY PRACTICES

The Good Laboratory Practices (GLPs) refers to a "standard" promulgated by the U.S. Food and Drug Administration (FDA) and the U.S. Environmental Protection Agency (EPA) to which all laboratories submitting scientific data to the agencies in support of applications for new drugs or medical devices, applications for registration, marketing permits, and the like, must conform. The choice of the name "good laboratory practices" was unfortunate, because chemists not familiar with the field sometimes interpret the phrase to simply mean good science, and are not aware that it refers to a detailed set of regulations on laboratory management that must be adhered to, to obtain government approval.

In the mid-1970s the U.S. Congress passed a series of laws concerning environmental protection and public health and safety, such as the Clean Drinking Water Act (CDWA), the Resource Conservation and Recovery Act (RCRA), the Federal Insecticide, Fungicide and Rodenticide Act (FIFRA), the Toxic Substances Control Act (TSCA), and the Occupational Safety and Health Act (OSHA) and others, which had the effect of an enormous increase in the demand for analytical services. The federal agencies charged with enforcing these laws did not have the personnel and facilities to respond to this demand, and were forced to rely on private sector laboratories for much of this work.

In the mid-1970s a series of widely publicized scandals indicated that the private sector could not be relied upon to provide quality data to meet the agencies' needs without a set of established standards. Hence, the FDA promulgated the first version of the GLPs in 1978. These were subsequently revised in 1987. These regulations applied to all data submitted to the agency for new drug applications and other regulatory purposes. The EPA followed in 1983 with its version of the GLPs for compliance to RCRA and FIFRA, which were only slightly different from those of the FDA. The FIFRA GLPs were revised in 1988.

In general the two agencies have tried to maintain consistency with each other as much as possible in view of their different missions. While both agencies are concerned with the toxicology of chemical substances, the FDA focuses on health effects and the EPA on environmental effects.

The GLPs are concerned chiefly with what are called "nonclinical" studies, that is, studies that do not involve experimentation on human subjects. Much of the content of the GLPs is not relevant to the analytical chemist, since it involves laboratory animal science, that is, care of laboratory animals, reporting results on animals, biological specimen handling, and so on. However, since much of the work of these studies also involves chemical analyses at some point of the protocol, the analytical laboratory manager must be aware of the requirements if he or she is asked to do work that will be part of a nonclinical study.

Such a request may come from either the "sponsor" of the study (the person or organization paying for it) or the "study director" who is the individual responsible for conducting the study. The sponsor of such a study is required by the regulations to inform laboratory management that the work must be done under the GLPs.

One difficulty with the GLPs is that some of the requirements are rather loosely defined, although others are rather rigid. For example, it is required that quality assurance audits be carried out to verify data integrity, on various phases or segments of the study. However, it is not clear what is meant by a phase or segment. Is the audit to be a system audit, a method audit, or an audit of just one step in the analytical process? For this reason it is good for the laboratory manager to discuss with the study director exactly what is necessary to meet the client's needs. Then reduce the points of agreement to writing to avoid possible future conflict or misinterpretation. Various departments and individuals in the EPA and FDA may also vary in their interpretation of the GLPs, and this should be taken into account before the work is done.

Reading the GLPs in their original form in the Code of Federal Regulations (CFR) can be an exercise in frustration for someone not familiar with the jargon used by lawyers in writing these laws, or with the peculiar system of headings, subheadings, paragraphs, subparagraphs, and so on used in this type of document. In what follows we will attempt to summarize the more important requirements of the GLPs as they affect the quality assurance program of the laboratory. This is not intended as a substitute for reading the regulations in detail, which must be done by the laboratory manager or his or her substitute before embarking on GLP work. The full text of the GLPs can be obtained from the relevant agency (FDA or EPA).

A quality assurance unit (QAU) must be provided. This is bureaucratese for a quality assurance officer and his or her staff. (The QAU may be a single person.)

The QAU must be separate from, and independent of, personnel engaged in the direction and conduct of the study.

Personnel engaged in the direction and conduct of the study must have the education, training, and experience necessary to perform their functions. Personnel training must be documented.

A written quality assurance manual must be provided, although the GLPs do not specify the contents of the manual.

Written standard operating procedures must be available for all operations, including analytical methods.

A historical file must be maintained of all SOPs.

Written calibration and maintenance procedures must be available, and all calibration and maintenance work must be documented.

Storage space for samples after receipt and prior to analysis must be provided, separate from the laboratory and containing necessary equipment such as refrigeration for sample preservation.

An archive area must be provided for storage of all relevant data and documents. The archive area must be of limited access.

Periodic audits or inspections of the various phases or segments of the study must be carried out by the QAU and documented.

Written sampling methods must be available if sampling is performed by laboratory personnel. (If not, this is the responsibility of the organization doing the sampling.)

All reagents and reagent solutions must be properly labeled with identity and concentration of components, date of preparation, storage requirements, and expiration date.

The GLPs contain many more requirements than these, but this list covers most of those pertaining to the analytical chemistry laboratory and its quality assurance program. It will be noted that all of the above items have been covered in previous chapters and it is safe to say that a quality assurance program as described in this book will satisfy the quality assurance requirements of the GLPs.

The GLPs as originally written are concerned mainly with studies that evaluate biological effects of chemical materials. However, the EPA is also concerned with environmental pollution, hazardous waste, and toxic materials handling and disposal. In these cases the GLPs have been extended beyond the requirements of the original documents. For example, in the EPA "Contract Laboratory Program" for analyzing materials from hazardous waste landfills, the contracting laboratory, in addition to adhering to the GLPs, must also use EPA-mandated sampling procedures, analytical procedures, and quality control procedures that are quite specific and detailed.

OTHER MANDATED QUALITY ASSURANCE PROGRAMS

In addition to the EPA and FDA many other government agencies have mandated quality assurance programs that must be complied with by laboratories under contract or representing organizations attempting to comply with regulations. For example, the U.S. Army

Toxic and Hazardous Materials Agency (USATHAMA) has a very comprehensive quality assurance program for contract laboratories doing work under the installation restoration projects program, which is concerned with identifying and cleaning up toxic and hazardous waste dumps at army installations. The U.S. Navy has a similar program.

At the state level most states have an environmental agency similar to the EPA, under names such as Department of Environmental Regulation, Department of Natural Resources, and the like. In many cases these agencies have assumed responsibilities for environmental testing under EPA supervision, with the usual stipulation being that the state program must be at least as rigorous as the EPAs. This has led to some very tight state quality assurance programs for laboratories doing analytical work for compliance to state regulations, especially in states where pollution is perceived to be a serious problem, for example, New Jersey, New York, and California. These regulations are often quite specific regarding analytical methods to be used and quality control measures.

Under the Clean Drinking Water Act, the EPA has a mandated quality assurance program, including specific analytical procedures to be used in analyzing drinking waters. In most states the accreditation or certification of laboratories that analyze drinking water is the responsibility of the state, but the state regulations must be at least as stringent as those of the EPA.

Municipal agencies, especially those involved in waste treatment and handling, may require laboratories to have a mandated quality assurance program. In many cases, the regulations covering the National Pollution Discharge Elimination System (NPDES) apply. Under these regulations the laboratory submitting data in support of a permit to discharge wastewater is required to analyze performance evaluation samples periodically and obtain acceptable results. There is no mandated quality assurance program, but should the laboratory fail to analyze the samples correctly, the laboratory will be inspected and required to exhibit its quality assurance procedures.

In many cases acceptable quality assurance programs are a most important factor in the awarding of government contracts, such as the contract laboratory program of the EPA for laboratories offering analytical services for the Superfund hazardous waste site program. Engineering companies, bidding on government contracts, may find that a laboratory with a tight quality assurance program is the deciding factor in obtaining the contract.

In the private sector many large companies are requiring their suppliers to demonstrate that analytical data produced in support of materials specifications are being generated under strict quality assurance programs.

LABORATORY ACCREDITATION

The proliferation of mandated quality assurance programs has led to a strong movement toward the establishment of third-party accreditation programs, in an attempt to alleviate the problems of laboratories forced to operate under a number of mandated programs that may differ in details, and indeed, sometimes be in conflict with each other. This problem is especially acute for the independent commercial testing laboratories that serve hundreds of clients with differing quality assurance needs. It would be desirable if the various organizations requiring mandated quality assurance programs could be persuaded to accept accreditation by an outside organization as equivalent to their own accreditation. Thus, one accreditation process could serve in place of many, and the result would be a great economic saving by the laboratories involved, which could be passed on to clients.

There is some confusion in the literature, and in the minds of many in the analytical laboratory community, regarding the use of the terms "accreditation" and "certification." Many persons use them interchangeably, and also mix them with the terms "registered," "authorized," "approved," and so on. In recent years, due to the efforts of international standards organizations, agreement seems to have been reached to use the term "accredited" to apply to organizations and "certified" to individuals or products. Thus, laboratories, schools, and hospitals may be accredited, while public accountants, industrial hygienists, and chemists (through the American Institute of Chemists) may be certified. This distinction has not been universally accepted; for example, the EPA still "certifies" laboratories for drinking water analysis.

There are two different philosophies regarding laboratory accreditation. One, often called product/standard accreditation, bases accreditation on a laboratory's competence to perform certain standard tests, usually on a restricted range of matrices or sample types.

For each analysis run or each group of similar analyses, a separate accreditation must be sought and competence demonstrated, although certain "generic" criteria such as quality assurance programs, facilities, or personnel training must also be satisfactory.

The other type of accreditation is called accreditation by scientific discipline or "field of testing." In this case the laboratory is accredited according to its competence in a given field, such as analytical chemistry, or in a group of subdisciplines, called testing technologies, such as gas chromatography, wet chemistry, or spectrophotometry. Accreditation is not given for specific tests on specific sample types, but rather on the laboratory's competence in the field as demonstrated by facilities, equipment, personnel training and experience, and quality assurance program.

These two different philosophies are demonstrated by the two major national accrediting agencies in the United States. These are the National Voluntary Laboratory Accreditation Program (NVLAP, pronounced "navlap"), and the American Association for Laboratory Accreditation (AALA, often designated A2LA).

The American Council of Independent Laboratories (ACIL) was one of the prime movers in the establishment of both accrediting agencies. ACIL is a trade association of independent, commercial testing laboratories of all types (not just analytical chemistry laboratories) and has promoted the notion of independent accrediting agencies since its beginning in the mid-1930s. High professional standards have been a hallmark of ACIL since its founding and its members felt that accreditation would prevent unfair competition by laboratories offering low-quality testing at lower prices to clients who were naive about the concept of laboratory quality assurance.

Accordingly, in the early 1970s ACIL and other trade organizations began to exert pressure on the federal government to establish a national program to accredit laboratories. Other countries had successfully established such agencies and a typical model, often cited, was the Australian National Association of Testing Authorities (NATA), which had a long and honorable record of over 30 years of laboratory accreditation. NATA is a field of testing type of accreditation, with some 10 fields of testing, such as chemical analysis, biological testing, electrical, and thermal. ACIL strongly supports the field of testing type of accreditation since many of its members run a wide variety of tests on a variety of sample types, and the product/standard type of accreditation would represent an unreasonable amount of work and cost for these laboratories.

In 1975 the U.S. Department of Commerce announced plans to establish a National Voluntary Laboratory Accreditation Program, which would be based on the field of testing approach. In 1976, after 10 months of study and public comment, the NVLAP was formally established, but on the product/standard approach, based presumably on the comments received.

This announcement was greeted with consternation by ACIL and other advocates of the field of testing approach. Realizing that the government was committed to the product/standard basis, these organizations decided to establish a nonprofit, private sector organization to accredit laboratories by field of testing. Thus, at a meeting in 1978 attended by representatives of 40 organizations, including ACIL, the American Association for Laboratory Accreditation was formed. A2LA is a nonprofit organization, funded by donations, membership dues, and accreditation fees. Although in general accreditation is by field of testing, in certain cases product/standard accreditation is offered. Examples are environmental chemistry, where strict adherence to EPA methods is required, and coal testing, where accreditation is on the basis of adherence to ASTM standards.

A2LA and NVLAP are the only organizations in the United States whose sole purpose is accrediting laboratories in a wide variety of scientific and engineering disciplines. There are other accrediting bodies, but they are usually limited to one type of testing, or exist to further the professional goals of a particular discipline. For example, the American Industrial Hygiene Institute (AIHA) accredits laboratories, but only those engaged in laboratory and sampling work for industrial hygiene, and only in those testing technologies relevant to this type of work. As other examples, the American Society of Crime Laboratory Directors accredits police laboratories, while the College of American Pathologists accredits hospital laboratories.

For accreditation in the broad field of analytical chemistry, with attention to specific testing technologies, the A2LA approach is probably most suitable. On the other hand if accreditation is required for only a limited number of specific tests, the NVLAP may prove suitable. Laboratories with a need for accreditation in a specific field of application, such as industrial hygiene, may be able to find accreditation programs offered by professional organizations in their field.

International Aspects of Laboratory Accreditation

In recent years the lack of an international system for laboratory accreditation has been perceived as a barrier to international commerce. If a commodity is traded between two countries it is often analyzed or tested in both the importing and exporting country, a duplication of effort that could be eliminated if the laboratories in both countries were demonstrably equivalent in competence.

In 1977, at the initiative of the United States and Denmark, the first meeting of the International Laboratory Accreditation Conference (ILAC) was held in Denmark. Since then ILAC has met every year to study the problems involved in obtaining international agreement on laboratory accreditation. ILAC is an informal assembly of government personnel and private sector organizations with an interest in the field. It has no official status, but acts as a forum for discussion of problems.

The work of ILAC is far from completed. It has published a directory of accrediting organizations, defined and discussed many of the legal problems of international recognition of accreditation, listed basic terms and definitions, and produced guidelines for quality assurance systems for laboratories and for accrediting organizations.

Criteria for Laboratory Accreditation

In most accreditation programs, the criteria for granting accreditation involve the following aspects of laboratory management:

Personnel. Personnel have the education, training, and experience to enable them to perform the tests for which accreditation is requested. Organization of personnel is such that authority over the technical operations and responsibility for quality is clearly defined.

Facilities. Laboratory facilities are adequate to perform the tests designated. Sufficient work space is provided, necessary utilities are available, and the laboratory environment is controlled as necessary for the work to be done.

Instrumentation and equipment. Instrumentation and equipment appropriate to the range of tests run are available, well maintained, and in good condition.

Quality assurance program. A suitable quality assurance program and necessary quality control procedures are in place and operating. In almost all accreditation programs a quality assurance manual is required.

The Accreditation Process

Most accrediting organizations follow a similar process in accrediting laboratories. The objective is to verify that the above criteria are satisfied in the day-to-day operations of the laboratory. The process usually involves the following steps:

1. The laboratory applies for accreditation to the accrediting organization. If laboratory management is not sure whether they wish to be accredited they may request information about the accrediting agency and the benefits of accreditation.

2. The laboratory will be sent an application form, to be filled out by the laboratory, and possibly a request for an application fee. The application form will usually request a great deal of information about the laboratory and its operations. The quality assurance officer or laboratory manager can expect to answer detailed questions regarding personnel and personnel organization; facilities, including floor plans; lists of instruments and equipment, including manufacturer, model number, and age; a copy of the laboratory quality assurance manual; description of the proficiency sample-testing programs the laboratory is engaged in; and other details. The laboratory will also be asked to specify in which fields of testing and testing technologies it would like to be accredited. Give the information requested as honestly as possible, since it will be verified later during the on-site assessment or inspection.

The information on the application will be carefully evaluated by the accrediting agency. It may happen that for one reason or another, the laboratory fails to meet the agency's criteria. If this is so, the laboratory will be so informed and given the option of correcting the deficiency or withdrawing the request for accreditation.

3. The next step will be to arrange for the on-site inspection. The accrediting agency will appoint one or more assessors, depending on the size of the laboratory and the number of testing technologies for which accreditation is requested. Laboratory assessors are generally experts in the fields covered by the accreditation and will be

carefully chosen to avoid any apparent or real conflict of interest (such as a former employee of the laboratory, or of a competitor, or a friend or relative of the laboratory owner or manager). At this point also, the laboratory will be given the opportunity of requesting another assessor if there is good and sufficient reason to reject one of the appointed assessors.

The assessors will contact the laboratory to arrange for a mutually convenient day or days for the inspection. Even a small laboratory will usually require at least one full day. Larger laboratories may require several days. The assessment must take place during normal operating hours, since the daily operations of the laboratory must be observed, and all essential personnel must be present. Since there will be a certain amount of disruption of laboratory work, inform all employees of the reason for the disruption, and also instruct them to answer all questions of the assessors as honestly and truthfully as possible.

4. On the appointed day the assessors will arrive in the morning, usually a half-hour or hour after the normal day begins to give employees a chance to accomplish any necessary early morning chores. The first order of business will be a meeting with laboratory management, the quality assurance officer, and any other key personnel management feels should be present. The assessors will explain the purpose of the inspection and the general agenda that will be followed.

The first step of the agenda will usually be a walk-through tour of the laboratory, accompanied by the laboratory manager or quality assurance officer. This will give the assessors a general impression of the laboratory's size, number of personnel, equipment, and facilities. The prudent assessor will also be looking for signs of professional attitude such as housekeeping and safety awareness.

The tour will be followed by the assessors' inspection, necessary to verify the laboratory's compliance to the criteria of the accrediting agency. Inspection of laboratory notebooks, records of all types, equipment, analytical methods, and all aspects of quality assurance will be scrutinized. The assessors will need to talk to technical personnel at all levels in the laboratory to assess their training, education, and attitude toward quality work. Warn laboratory supervisory staff of this and have them instruct their subordinates accordingly. Warn supervisors to resist the temptation to answer for their subordinates.

One common technique the assessor may use is to request a

random analytical report, and ask the supervisor or quality assurance officer to trace the results back to the raw data, analyst, equipment, date, and relevant quality control charts. Another technique is to follow an incoming sample through the steps of log-in, storage, assignment to analyst, and so on.

The assessors will inspect for compliance to the "generic criteria" (see the checklist in Chapter 10), specific criteria for the various testing technologies, and individual test criteria, if necessary. Checklists, such as that given in Chapter 10, are usually used, but the inspection is not necessarily confined only to items on the checklists.

The inspectors will then request an office or quiet desk somewhere to prepare their report. In most cases the assessors will then present their findings to laboratory management before leaving the laboratory, with a (usually) hand-written report.

The report will cover two different categories of observations. The first will be "deficiencies," that is, conditions or operations that preclude the laboratory from being accredited because they illustrate lack of compliance to the accrediting organization's criteria. Laboratory management will have the opportunity to explain or clear up what may be a misconception on the part of the assessor. The assessor may or may not amend the report. Laboratory management will be asked to sign the report, indicating that it was presented to them and read, whether or not they agree with the findings.

The assessors cannot accredit the laboratory. This is the function of the accrediting organization; however, the organization will rely heavily on the assessor's report. If accreditation is denied the laboratory has two options. They may appeal the assessor's report and ask for reconsideration, which will be done according to the regulations of the accrediting agency, or they may be given a period of time (usually 60 to 90 days) to correct the deficiencies found. Documentary proof of the corrections may be required.

The other category of assessor observations might be called simply "recommendations." These are not sufficiently serious to deny accreditation, but represent possible improvements that could be made in the various aspects of laboratory management, including aspects not related to laboratory accreditation. Accreditation assessors are generally experienced in both technical aspects of analytical chemistry and in laboratory management, and their recommendations should be carefully considered. Often an outside observer can find weaknesses in laboratory management that are not readily apparent to those in the organization. This is one of the side benefits of the accreditation process.

5. In addition to the on-site assessment, the accrediting body may require the analysis of performance evaluation samples. These may be supplied by the agency, or the laboratory may be required to participate in a performance evaluation program such as the EPA or state drinking water program, or the NIOSH performance analytical test (PAT) program. Failure to analyze samples acceptably may result in denial of accreditation, and the laboratory may be requested to analyze another set of samples and convincingly explain the unacceptable results.

6. After successful analysis of the performance evaluation samples and correction of any deficiencies found in the assessors' report, the laboratory will be accredited, pending the payment of the required fees.

However, accreditation is an on-going process, and the laboratory will be required to maintain the necessary standards to retain accreditation. On-site inspections will be periodically scheduled, usually every two years. Periodic analysis of performance evaluation samples will be required as well. In addition the laboratory will be required to keep the accrediting agency informed of any changes that could affect the quality of the work, such as changes in the status of key personnel, changes in personnel organization, new instrumentation purchases, results of other performance evaluation programs, and litigation involving quality of analytical data.

Costs and Benefits of Accreditation

Achieving accreditation can be a costly process. In addition to the fees charged by the accrediting agency, the laboratory will be charged with the fees for the laboratory assessors, as well as their travel and living expenses during the assessment. Although some of these costs are variable, depending on the laboratory's size, many of them are fixed, such as the basic administrative fee charged by the agency. The result can be costly, especially for the smaller laboratory operating on a small budget.

The laboratory manager must thus weigh the cost of accreditation carefully against the benefits. In many cases, there is no option: accreditation is necessary for the laboratory to fulfill its mission. In other cases the benefits may not be so apparent, but may nevertheless be quite real. Certainly, assessment of a laboratory by a disinterested third party will usually improve laboratory performance by showing up areas of vulnerability or weaknesses in operations, which may not be readily apparent to the laboratory supervisors or quality assurance officer. Achieving accreditation in a rigorous program will also be found to have a beneficial effect on the morale of laboratory workers since they now have proof of the fact that they are producing quality data.

REFERENCE

1. Hess, Earl H. Laboratory Accreditation, Professionalism Reduced to Practice, Paper presented at the annual meeting of the Texas Council of Engineering Laboratories, Fort Worth, Texas, 1984.

Appendix

Table
A-1. Critical Values of the *F* Test (95% confidence values)

	df_N									
df_D	1	2	3	4	5	6	7	8	9	10
1	648	800	864	900	922	937	948	956	963	968
2	38.5	39.0	39.2	29.3	39.3	39.4	39.4	39.4	39.4	39.4
3	17.4	16.0	15.4	15.1	14.9	14.7	14.6	14.5	14.5	14.4
4	12.2	10.6	9.98	9.60	9.36	9.20	9.07	8.98	8.90	8.84
5	10.0	8.43	7.76	7.39	7.14	6.98	6.85	6.75	6.68	6.62
6	8.81	7.26	6.60	6.23	5.99	5.82	5.70	5.60	5.52	5.46
7	8.07	6.54	5.89	5.52	5.28	5.12	4.99	4.90	4.82	4.76
8	7.57	6.06	5.42	5.05	4.82	4.65	4.53	4.43	4.36	4.29
9	7.21	5.71	5.08	4.72	4.48	4.32	4.20	4.10	4.03	3.96
10	6.94	5.46	4.82	4.47	4.24	4.07	3.95	3.85	3.78	3.72

Note: df_N is the degrees of freedom in the numerator, and df_D is the degrees of freedom in the denominator of *F*.

Table
A-2. Critical Values of the *F* Test (99% confidence values)

df_D	df_N 1	2	3	4	5	6	7	8	9
1	16,211	20,000	21,615	22,500	23,056	23,437	23,715	23,925	24,091
2	198.5	199.0	199.2	199.3	199.3	199.3	199.4	199.4	199.4
3	55.5	49.8	47.5	46.2	45.4	44.8	44.4	44.1	43.9
4	31.3	26.3	24.3	22.2	22.5	22.0	21.6	21.4	21.1
5	22.8	18.3	16.5	15.6	14.9	14.5	14.2	14.0	13.8
6	18.6	14.5	12.9	12.0	11.5	11.1	10.8	10.6	10.4
7	16.2	12.4	10.9	10.0	9.52	9.15	8.89	8.67	8.51
8	14.7	11.0	9.60	8.80	8.30	7.95	7.69	7.50	7.34
9	13.6	10.1	8.72	7.96	7.47	7.13	6.88	6.69	6.54
10	12.8	9.43	8.08	7.34	6.87	6.54	6.30	6.12	5.97

Note: df_N is the degrees of freedom in the numerator, and df_D is the degrees of freedom in the denominator of *F*.

Additional Readings

Journal Articles

Bankston, D. 1984. Elimination of Bias in Sample Selection for Chemical Analysis, *American Laboratory*, pp. 43–48 (December 1984).

Boomer, B., T. Dux, and D. March. Sampling Surveys of Hazardous Waste, *Journal of Air Pollution Control and Waste Management*, 38:1426–1432 (1988).

Dols, T., and B. Armbrecht. Assessment of Analytical Method Performance Characteristics, *Journal of the AOAC*, 60:940–945 (1977).

Dux, J. P. Quality Assurance in the Analytical Laboratory, *American Laboratory*, pp. 212–216 (July 1983).

Glacer, J. L. Trace Analysis for Wastewater, *Environmental Science and Technology*, 15:1426–1455 (1981).

Horwitz, W., L. R. Kampes, and K. W. Boyer. Quality Assurance in the Analysis of Foods for Trace Constituents, *Journal of the Association of Official Analytical Chemists*, 63:1344–1354 (1980).

Hubaux, A., and G. Vos. Precision and Detection Limits for Linear Calibration Curves, *Analytical Chemistry*, 42:849–855 (1970).

Kanzelmyer, J. H. Quality Control for Analytical Methods, *ASTM Standardization News*, pp. 25–28 (October 1977).

Keith, L. H. et al. Principles of Environmental Analysis, *Analytical Chemistry*, 55:2210–2218 (1983).

Kirchmer, C. J. Quality Control in Water Analysis, *Environmental Science and Technology*, 17:174A–184A (1983).

Kratchovil, B., D. Wallace, and J. Taylor. Sampling for Chemical Analysis, *Chemical Reviews*, 56:113R (1984).

MacDougall, D. et al. Guidelines for Data Acquisition and Data Quality Evaluation in Environmental Chemistry, *Analytical Chemistry*, 52:2242–2249 (1980).

Mandel, J. The Analysis of Interlaboratory Test Data, *ASTM Standardization News*, pp. 17–20 (1977).

Mandel, J., and F. C. Linnig. Study of Accuracy in Chemical Analyses Using Linear Calibration Curves, *Analytical Chemistry*, 29:743–749 (1959).

Provost, L. P., and R. S. Elder. Interpretation of Percent Recovery Data, *American Laboratory*, pp. 57–63 (1983).

Saltzman, B. E., D. W. Yeager, and B. G. Meiners. Reproducibility and Quality Control in the Analysis of Biological Samples for Lead and Mercury, *American Industrial Hygiene Association Journal*, 44:163–167 (1983).

Schwartz, L. M. Nonlinear Calibration, *Analytical Chemistry*, 49:2062–2068 (1977).

Youden, W. Accuracy of Analytical Procedures, *Journal of the Association of Official Analytical Chemists*, 45:169–173 (1962).

Books

Beizer, P. *The Frozen Keyboard: Living with Bad Software*, Blue Ridge Summit, Pa.: TAB Books, Inc., 1988.

Dessy, R. *The Electronic Laboratory*, Washington, D.C.: American Chemical Society, 1985.

Dowdey, S., and S. Wearden. *Statistics for Research*. New York: John Wiley and Sons, 1983.

Fites, P., P. Johnston, and M. Kratz. *The Computer Virus Crisis*, New York: Van Nostrand Reinhold, 1989.

Garfield, F. M. *Quality Assurance Principles for Analytical Laboratories*, Arlington, Va.: Association of Official Analytical Chemists, 1984.

Hurst, W., and J. Mortimer. *Laboratory Robotics*, New York: VCH Publishers, 1987.

Inhorn, S. L. (ed.) *Quality Assurance Practices for Health Laboratories*, Washington, D.C.: American Public Health Association, 1978.

Kanare, H. W. *Writing the Laboratory Notebook*, Washington, D.C.: American Chemical Society, 1985.

Mahaffey, R. *LIMS: Applied Information Technology for the Laboratory*, New York: Van Nostrand Reinhold, 1990.

Schulmeyer, G., and J. McManus (eds.) *Handbook of Software Quality Assurance*, New York: Van Nostrand Reinhold, 1987.

Taylor, J. *Quality Assurance of Chemical Measurements*, Chelsea, Mich.: Lewis Publishers, 1988.

Youden, W. J., and E. H. Steiner. *Statistical Manual of the AOAC*, Washington, D.C.: Association of Official Analytical Chemists, 1975.

INDEX

GPSR Compliance
The European Union's (EU) General Product Safety Regulation (GPSR) is a set of rules that requires consumer products to be safe and our obligations to ensure this.

If you have any concerns about our products, you can contact us on

ProductSafety@springernature.com

In case Publisher is established outside the EU, the EU authorized representative is:

Springer Nature Customer Service Center GmbH
Europaplatz 3
69115 Heidelberg, Germany

www.ingramcontent.com/pod-product-compliance
Ingram Content Group UK Ltd.
Pitfield, Milton Keynes, MK11 3LW, UK
UKHW041334110926
15112UKWH00015B/39
* 9 7 8 1 4 7 5 7 5 3 7 8 3 *